Energy Systems: A Very Short Introduction

VERY SHORT INTRODUCTIONS are for anyone wanting a stimulating and accessible way into a new subject. They are written by experts, and have been translated into more than 45 different languages.

The series began in 1995, and now covers a wide variety of topics in every discipline. The VSI library currently contains over 600 volumes—a Very Short Introduction to everything from Psychology and Philosophy of Science to American History and Relativity—and continues to grow in every subject area.

Very Short Introductions available now:

ABOLITIONISM Richard S. Newman
ACCOUNTING Christopher Nobes
ADAM SMITH Christopher J. Berry
ADOLESCENCE Peter K. Smith
ADVERTISING Winston Fletcher
AESTHETICS Bence Nanay
AFRICAN AMERICAN RELIGION
   Eddie S. Glaude Jr
AFRICAN HISTORY John Parker and
   Richard Rathbone
AFRICAN POLITICS Ian Taylor
AFRICAN RELIGIONS
   Jacob K. Olupona
AGEING Nancy A. Pachana
AGNOSTICISM Robin Le Poidevin
AGRICULTURE Paul Brassley and
   Richard Soffe
ALEXANDER THE GREAT
   Hugh Bowden
ALGEBRA Peter M. Higgins
AMERICAN CULTURAL
   HISTORY Eric Avila
AMERICAN FOREIGN
   RELATIONS Andrew Preston
AMERICAN HISTORY
   Paul S. Boyer
AMERICAN IMMIGRATION
   David A. Gerber
AMERICAN LEGAL HISTORY
   G. Edward White
AMERICAN NAVAL HISTORY
   Craig L. Symonds
AMERICAN POLITICAL HISTORY
   Donald Critchlow

AMERICAN POLITICAL PARTIES
   AND ELECTIONS L. Sandy Maisel
AMERICAN POLITICS
   Richard M. Valelly
THE AMERICAN PRESIDENCY
   Charles O. Jones
THE AMERICAN REVOLUTION
   Robert J. Allison
AMERICAN SLAVERY
   Heather Andrea Williams
THE AMERICAN WEST Stephen Aron
AMERICAN WOMEN'S HISTORY
   Susan Ware
ANAESTHESIA Aidan O'Donnell
ANALYTIC PHILOSOPHY
   Michael Beaney
ANARCHISM Colin Ward
ANCIENT ASSYRIA Karen Radner
ANCIENT EGYPT Ian Shaw
ANCIENT EGYPTIAN ART AND
   ARCHITECTURE Christina Riggs
ANCIENT GREECE Paul Cartledge
THE ANCIENT NEAR EAST
   Amanda H. Podany
ANCIENT PHILOSOPHY Julia Annas
ANCIENT WARFARE
   Harry Sidebottom
ANGELS David Albert Jones
ANGLICANISM Mark Chapman
THE ANGLO-SAXON AGE John Blair
ANIMAL BEHAVIOUR
   Tristram D. Wyatt
THE ANIMAL KINGDOM
   Peter Holland

ANIMAL RIGHTS David DeGrazia
THE ANTARCTIC Klaus Dodds
ANTHROPOCENE Erle C. Ellis
ANTISEMITISM Steven Beller
ANXIETY Daniel Freeman and
    Jason Freeman
THE APOCRYPHAL GOSPELS
    Paul Foster
APPLIED MATHEMATICS
    Alain Goriely
ARCHAEOLOGY Paul Bahn
ARCHITECTURE Andrew Ballantyne
ARISTOCRACY William Doyle
ARISTOTLE Jonathan Barnes
ART HISTORY Dana Arnold
ART THEORY Cynthia Freeland
ARTIFICIAL INTELLIGENCE
    Margaret A. Boden
ASIAN AMERICAN
    HISTORY Madeline Y. Hsu
ASTROBIOLOGY David C. Catling
ASTROPHYSICS James Binney
ATHEISM Julian Baggini
THE ATMOSPHERE Paul I. Palmer
AUGUSTINE Henry Chadwick
AUSTRALIA Kenneth Morgan
AUTISM Uta Frith
AUTOBIOGRAPHY Laura Marcus
THE AVANT GARDE David Cottington
THE AZTECS David Carrasco
BABYLONIA Trevor Bryce
BACTERIA Sebastian G. B. Amyes
BANKING John Goddard and
    John O. S. Wilson
BARTHES Jonathan Culler
THE BEATS David Sterritt
BEAUTY Roger Scruton
BEHAVIOURAL
    ECONOMICS Michelle Baddeley
BESTSELLERS John Sutherland
THE BIBLE John Riches
BIBLICAL ARCHAEOLOGY
    Eric H. Cline
BIG DATA Dawn E. Holmes
BIOGRAPHY Hermione Lee
BIOMETRICS Michael Fairhurst
BLACK HOLES Katherine Blundell
BLOOD Chris Cooper
THE BLUES Elijah Wald
THE BODY Chris Shilling

THE BOOK OF COMMON PRAYER
    Brian Cummings
THE BOOK OF MORMON
    Terryl Givens
BORDERS Alexander C. Diener and
    Joshua Hagen
THE BRAIN Michael O'Shea
BRANDING Robert Jones
THE BRICS Andrew F. Cooper
THE BRITISH CONSTITUTION
    Martin Loughlin
THE BRITISH EMPIRE Ashley Jackson
BRITISH POLITICS Anthony Wright
BUDDHA Michael Carrithers
BUDDHISM Damien Keown
BUDDHIST ETHICS Damien Keown
BYZANTIUM Peter Sarris
C. S. LEWIS James Como
CALVINISM Jon Balserak
CANCER Nicholas James
CAPITALISM James Fulcher
CATHOLICISM Gerald O'Collins
CAUSATION Stephen Mumford and
    Rani Lill Anjum
THE CELL Terence Allen and
    Graham Cowling
THE CELTS Barry Cunliffe
CHAOS Leonard Smith
CHARLES DICKENS Jenny Hartley
CHEMISTRY Peter Atkins
CHILD PSYCHOLOGY Usha Goswami
CHILDREN'S LITERATURE
    Kimberley Reynolds
CHINESE LITERATURE Sabina Knight
CHOICE THEORY Michael Allingham
CHRISTIAN ART Beth Williamson
CHRISTIAN ETHICS D. Stephen Long
CHRISTIANITY Linda Woodhead
CIRCADIAN RHYTHMS
    Russell Foster and Leon Kreitzman
CITIZENSHIP Richard Bellamy
CIVIL ENGINEERING
    David Muir Wood
CLASSICAL LITERATURE William Allan
CLASSICAL MYTHOLOGY
    Helen Morales
CLASSICS Mary Beard and
    John Henderson
CLAUSEWITZ Michael Howard
CLIMATE Mark Maslin

CLIMATE CHANGE Mark Maslin
CLINICAL PSYCHOLOGY
    Susan Llewelyn and
    Katie Aafjes-van Doorn
COGNITIVE NEUROSCIENCE
    Richard Passingham
THE COLD WAR Robert McMahon
COLONIAL AMERICA Alan Taylor
COLONIAL LATIN AMERICAN
    LITERATURE Rolena Adorno
COMBINATORICS Robin Wilson
COMEDY Matthew Bevis
COMMUNISM Leslie Holmes
COMPARATIVE LITERATURE
    Ben Hutchinson
COMPLEXITY John H. Holland
THE COMPUTER Darrel Ince
COMPUTER SCIENCE
    Subrata Dasgupta
CONCENTRATION CAMPS
    Dan Stone
CONFUCIANISM Daniel K. Gardner
THE CONQUISTADORS
    Matthew Restall and
    Felipe Fernández-Armesto
CONSCIENCE Paul Strohm
CONSCIOUSNESS Susan Blackmore
CONTEMPORARY ART
    Julian Stallabrass
CONTEMPORARY FICTION
    Robert Eaglestone
CONTINENTAL PHILOSOPHY
    Simon Critchley
COPERNICUS Owen Gingerich
CORAL REEFS Charles Sheppard
CORPORATE SOCIAL
    RESPONSIBILITY Jeremy Moon
CORRUPTION Leslie Holmes
COSMOLOGY Peter Coles
COUNTRY MUSIC Richard Carlin
CRIME FICTION Richard Bradford
CRIMINAL JUSTICE Julian V. Roberts
CRIMINOLOGY Tim Newburn
CRITICAL THEORY
    Stephen Eric Bronner
THE CRUSADES Christopher Tyerman
CRYPTOGRAPHY Fred Piper and
    Sean Murphy
CRYSTALLOGRAPHY A. M. Glazer
THE CULTURAL REVOLUTION
    Richard Curt Kraus

DADA AND SURREALISM
    David Hopkins
DANTE Peter Hainsworth and
    David Robey
DARWIN Jonathan Howard
THE DEAD SEA SCROLLS
    Timothy H. Lim
DECADENCE David Weir
DECOLONIZATION Dane Kennedy
DEMOCRACY Bernard Crick
DEMOGRAPHY Sarah Harper
DEPRESSION Jan Scott and
    Mary Jane Tacchi
DERRIDA Simon Glendinning
DESCARTES Tom Sorell
DESERTS Nick Middleton
DESIGN John Heskett
DEVELOPMENT Ian Goldin
DEVELOPMENTAL BIOLOGY
    Lewis Wolpert
THE DEVIL Darren Oldridge
DIASPORA Kevin Kenny
DICTIONARIES Lynda Mugglestone
DINOSAURS David Norman
DIPLOMACY Joseph M. Siracusa
DOCUMENTARY FILM
    Patricia Aufderheide
DREAMING J. Allan Hobson
DRUGS Les Iversen
DRUIDS Barry Cunliffe
DYNASTY Jeroen Duindam
DYSLEXIA Margaret J. Snowling
EARLY MUSIC Thomas Forrest Kelly
THE EARTH Martin Redfern
EARTH SYSTEM SCIENCE Tim Lenton
ECONOMICS Partha Dasgupta
EDUCATION Gary Thomas
EGYPTIAN MYTH Geraldine Pinch
EIGHTEENTH-CENTURY BRITAIN
    Paul Langford
THE ELEMENTS Philip Ball
EMOTION Dylan Evans
EMPIRE Stephen Howe
ENERGY SYSTEMS Nick Jenkins
ENGELS Terrell Carver
ENGINEERING David Blockley
THE ENGLISH LANGUAGE
    Simon Horobin
ENGLISH LITERATURE Jonathan Bate
THE ENLIGHTENMENT
    John Robertson

ENTREPRENEURSHIP Paul Westhead
    and Mike Wright
ENVIRONMENTAL
    ECONOMICS Stephen Smith
ENVIRONMENTAL ETHICS
    Robin Attfield
ENVIRONMENTAL LAW
    Elizabeth Fisher
ENVIRONMENTAL POLITICS
    Andrew Dobson
EPICUREANISM Catherine Wilson
EPIDEMIOLOGY Rodolfo Saracci
ETHICS Simon Blackburn
ETHNOMUSICOLOGY Timothy Rice
THE ETRUSCANS Christopher Smith
EUGENICS Philippa Levine
THE EUROPEAN UNION
    Simon Usherwood and John Pinder
EUROPEAN UNION LAW
    Anthony Arnull
EVOLUTION Brian and
    Deborah Charlesworth
EXISTENTIALISM Thomas Flynn
EXPLORATION Stewart A. Weaver
EXTINCTION Paul B. Wignall
THE EYE Michael Land
FAIRY TALE Marina Warner
FAMILY LAW Jonathan Herring
FASCISM Kevin Passmore
FASHION Rebecca Arnold
FEDERALISM Mark J. Rozell and
    Clyde Wilcox
FEMINISM Margaret Walters
FILM Michael Wood
FILM MUSIC Kathryn Kalinak
FILM NOIR James Naremore
THE FIRST WORLD WAR
    Michael Howard
FOLK MUSIC Mark Slobin
FOOD John Krebs
FORENSIC PSYCHOLOGY
    David Canter
FORENSIC SCIENCE Jim Fraser
FORESTS Jaboury Ghazoul
FOSSILS Keith Thomson
FOUCAULT Gary Gutting
THE FOUNDING FATHERS
    R. B. Bernstein
FRACTALS Kenneth Falconer
FREE SPEECH Nigel Warburton
FREE WILL Thomas Pink

FREEMASONRY Andreas Önnerfors
FRENCH LITERATURE John D. Lyons
THE FRENCH REVOLUTION
    William Doyle
FREUD Anthony Storr
FUNDAMENTALISM Malise Ruthven
FUNGI Nicholas P. Money
THE FUTURE Jennifer M. Gidley
GALAXIES John Gribbin
GALILEO Stillman Drake
GAME THEORY Ken Binmore
GANDHI Bhikhu Parekh
GARDEN HISTORY Gordon Campbell
GENES Jonathan Slack
GENIUS Andrew Robinson
GENOMICS John Archibald
GEOFFREY CHAUCER David Wallace
GEOGRAPHY John Matthews and
    David Herbert
GEOLOGY Jan Zalasiewicz
GEOPHYSICS William Lowrie
GEOPOLITICS Klaus Dodds
GERMAN LITERATURE Nicholas Boyle
GERMAN PHILOSOPHY
    Andrew Bowie
GLACIATION David J. A. Evans
GLOBAL CATASTROPHES Bill McGuire
GLOBAL ECONOMIC HISTORY
    Robert C. Allen
GLOBALIZATION Manfred Steger
GOD John Bowker
GOETHE Ritchie Robertson
THE GOTHIC Nick Groom
GOVERNANCE Mark Bevir
GRAVITY Timothy Clifton
THE GREAT DEPRESSION AND
    THE NEW DEAL Eric Rauchway
HABERMAS James Gordon Finlayson
THE HABSBURG EMPIRE
    Martyn Rady
HAPPINESS Daniel M. Haybron
THE HARLEM RENAISSANCE
    Cheryl A. Wall
THE HEBREW BIBLE AS LITERATURE
    Tod Linafelt
HEGEL Peter Singer
HEIDEGGER Michael Inwood
THE HELLENISTIC AGE
    Peter Thonemann
HEREDITY John Waller
HERMENEUTICS Jens Zimmermann

HERODOTUS Jennifer T. Roberts
HIEROGLYPHS Penelope Wilson
HINDUISM Kim Knott
HISTORY John H. Arnold
THE HISTORY OF ASTRONOMY
    Michael Hoskin
THE HISTORY OF CHEMISTRY
    William H. Brock
THE HISTORY OF CHILDHOOD
    James Marten
THE HISTORY OF CINEMA
    Geoffrey Nowell-Smith
THE HISTORY OF LIFE
    Michael Benton
THE HISTORY OF MATHEMATICS
    Jacqueline Stedall
THE HISTORY OF MEDICINE
    William Bynum
THE HISTORY OF PHYSICS
    J. L. Heilbron
THE HISTORY OF TIME
    Leofranc Holford-Strevens
HIV AND AIDS Alan Whiteside
HOBBES Richard Tuck
HOLLYWOOD Peter Decherney
THE HOLY ROMAN EMPIRE
    Joachim Whaley
HOME Michael Allen Fox
HOMER Barbara Graziosi
HORMONES Martin Luck
HUMAN ANATOMY
    Leslie Klenerman
HUMAN EVOLUTION Bernard Wood
HUMAN RIGHTS Andrew Clapham
HUMANISM Stephen Law
HUME A. J. Ayer
HUMOUR Noël Carroll
THE ICE AGE Jamie Woodward
IDENTITY Florian Coulmas
IDEOLOGY Michael Freeden
THE IMMUNE SYSTEM
    Paul Klenerman
INDIAN CINEMA
    Ashish Rajadhyaksha
INDIAN PHILOSOPHY Sue Hamilton
THE INDUSTRIAL REVOLUTION
    Robert C. Allen
INFECTIOUS DISEASE Marta L. Wayne
    and Benjamin M. Bolker
INFINITY Ian Stewart

INFORMATION Luciano Floridi
INNOVATION Mark Dodgson and
    David Gann
INTELLECTUAL PROPERTY
    Siva Vaidhyanathan
INTELLIGENCE Ian J. Deary
INTERNATIONAL LAW
    Vaughan Lowe
INTERNATIONAL
    MIGRATION Khalid Koser
INTERNATIONAL RELATIONS
    Paul Wilkinson
INTERNATIONAL SECURITY
    Christopher S. Browning
IRAN Ali M. Ansari
ISLAM Malise Ruthven
ISLAMIC HISTORY Adam Silverstein
ISOTOPES Rob Ellam
ITALIAN LITERATURE
    Peter Hainsworth and David Robey
JESUS Richard Bauckham
JEWISH HISTORY David N. Myers
JOURNALISM Ian Hargreaves
JUDAISM Norman Solomon
JUNG Anthony Stevens
KABBALAH Joseph Dan
KAFKA Ritchie Robertson
KANT Roger Scruton
KEYNES Robert Skidelsky
KIERKEGAARD Patrick Gardiner
KNOWLEDGE Jennifer Nagel
THE KORAN Michael Cook
LAKES Warwick F. Vincent
LANDSCAPE ARCHITECTURE
    Ian H. Thompson
LANDSCAPES AND
    GEOMORPHOLOGY
    Andrew Goudie and Heather Viles
LANGUAGES Stephen R. Anderson
LATE ANTIQUITY Gillian Clark
LAW Raymond Wacks
THE LAWS OF THERMODYNAMICS
    Peter Atkins
LEADERSHIP Keith Grint
LEARNING Mark Haselgrove
LEIBNIZ Maria Rosa Antognazza
LEO TOLSTOY Liza Knapp
LIBERALISM Michael Freeden
LIGHT Ian Walmsley
LINCOLN Allen C. Guelzo

LINGUISTICS Peter Matthews
LITERARY THEORY Jonathan Culler
LOCKE John Dunn
LOGIC Graham Priest
LOVE Ronald de Sousa
MACHIAVELLI Quentin Skinner
MADNESS Andrew Scull
MAGIC Owen Davies
MAGNA CARTA Nicholas Vincent
MAGNETISM Stephen Blundell
MALTHUS Donald Winch
MAMMALS T. S. Kemp
MANAGEMENT John Hendry
MAO Delia Davin
MARINE BIOLOGY Philip V. Mladenov
THE MARQUIS DE SADE John Phillips
MARTIN LUTHER Scott H. Hendrix
MARTYRDOM Jolyon Mitchell
MARX Peter Singer
MATERIALS Christopher Hall
MATHEMATICAL FINANCE
    Mark H. A. Davis
MATHEMATICS Timothy Gowers
MATTER Geoff Cottrell
THE MEANING OF LIFE
    Terry Eagleton
MEASUREMENT David Hand
MEDICAL ETHICS Michael Dunn and
    Tony Hope
MEDICAL LAW Charles Foster
MEDIEVAL BRITAIN John Gillingham
    and Ralph A. Griffiths
MEDIEVAL LITERATURE
    Elaine Treharne
MEDIEVAL PHILOSOPHY
    John Marenbon
MEMORY Jonathan K. Foster
METAPHYSICS Stephen Mumford
METHODISM William J. Abraham
THE MEXICAN REVOLUTION
    Alan Knight
MICHAEL FARADAY
    Frank A. J. L. James
MICROBIOLOGY Nicholas P. Money
MICROECONOMICS Avinash Dixit
MICROSCOPY Terence Allen
THE MIDDLE AGES Miri Rubin
MILITARY JUSTICE Eugene R. Fidell
MILITARY STRATEGY
    Antulio J. Echevarria II

MINERALS David Vaughan
MIRACLES Yujin Nagasawa
MODERN ARCHITECTURE
    Adam Sharr
MODERN ART David Cottington
MODERN CHINA Rana Mitter
MODERN DRAMA
    Kirsten E. Shepherd-Barr
MODERN FRANCE
    Vanessa R. Schwartz
MODERN INDIA Craig Jeffrey
MODERN IRELAND Senia Pašeta
MODERN ITALY Anna Cento Bull
MODERN JAPAN
    Christopher Goto-Jones
MODERN LATIN AMERICAN
    LITERATURE
    Roberto González Echevarría
MODERN WAR Richard English
MODERNISM Christopher Butler
MOLECULAR BIOLOGY Aysha Divan
    and Janice A. Royds
MOLECULES Philip Ball
MONASTICISM Stephen J. Davis
THE MONGOLS Morris Rossabi
MOONS David A. Rothery
MORMONISM Richard Lyman Bushman
MOUNTAINS Martin F. Price
MUHAMMAD Jonathan A. C. Brown
MULTICULTURALISM Ali Rattansi
MULTILINGUALISM John C. Maher
MUSIC Nicholas Cook
MYTH Robert A. Segal
NAPOLEON David Bell
THE NAPOLEONIC WARS
    Mike Rapport
NATIONALISM Steven Grosby
NATIVE AMERICAN
    LITERATURE Sean Teuton
NAVIGATION Jim Bennett
NAZI GERMANY Jane Caplan
NELSON MANDELA Elleke Boehmer
NEOLIBERALISM Manfred Steger and
    Ravi Roy
NETWORKS Guido Caldarelli and
    Michele Catanzaro
THE NEW TESTAMENT
    Luke Timothy Johnson
THE NEW TESTAMENT AS
    LITERATURE Kyle Keefer

NEWTON Robert Iliffe
NIETZSCHE Michael Tanner
NINETEENTH-CENTURY BRITAIN
Christopher Harvie and
H. C. G. Matthew
THE NORMAN CONQUEST
George Garnett
NORTH AMERICAN INDIANS
Theda Perdue and Michael D. Green
NORTHERN IRELAND
Marc Mulholland
NOTHING Frank Close
NUCLEAR PHYSICS Frank Close
NUCLEAR POWER Maxwell Irvine
NUCLEAR WEAPONS
Joseph M. Siracusa
NUMBERS Peter M. Higgins
NUTRITION David A. Bender
OBJECTIVITY Stephen Gaukroger
OCEANS Dorrik Stow
THE OLD TESTAMENT
Michael D. Coogan
THE ORCHESTRA D. Kern Holoman
ORGANIC CHEMISTRY
Graham Patrick
ORGANIZATIONS Mary Jo Hatch
ORGANIZED CRIME
Georgios A. Antonopoulos and
Georgios Papanicolaou
ORTHODOX CHRISTIANITY
A. Edward Siecienski
PAGANISM Owen Davies
PAIN Rob Boddice
THE PALESTINIAN-ISRAELI
CONFLICT Martin Bunton
PANDEMICS Christian W. McMillen
PARTICLE PHYSICS Frank Close
PAUL E. P. Sanders
PEACE Oliver P. Richmond
PENTECOSTALISM William K. Kay
PERCEPTION Brian Rogers
THE PERIODIC TABLE
Eric R. Scerri
PHILOSOPHY Edward Craig
PHILOSOPHY IN THE ISLAMIC
WORLD Peter Adamson
PHILOSOPHY OF BIOLOGY
Samir Okasha
PHILOSOPHY OF LAW
Raymond Wacks

PHILOSOPHY OF SCIENCE
Samir Okasha
PHILOSOPHY OF RELIGION
Tim Bayne
PHOTOGRAPHY Steve Edwards
PHYSICAL CHEMISTRY Peter Atkins
PHYSICS Sidney Perkowitz
PILGRIMAGE Ian Reader
PLAGUE Paul Slack
PLANETS David A. Rothery
PLANTS Timothy Walker
PLATE TECTONICS Peter Molnar
PLATO Julia Annas
POETRY Bernard O'Donoghue
POLITICAL PHILOSOPHY
David Miller
POLITICS Kenneth Minogue
POPULISM Cas Mudde and
Cristóbal Rovira Kaltwasser
POSTCOLONIALISM Robert Young
POSTMODERNISM Christopher Butler
POSTSTRUCTURALISM
Catherine Belsey
POVERTY Philip N. Jefferson
PREHISTORY Chris Gosden
PRESOCRATIC PHILOSOPHY
Catherine Osborne
PRIVACY Raymond Wacks
PROBABILITY John Haigh
PROGRESSIVISM Walter Nugent
PROJECTS Andrew Davies
PROTESTANTISM Mark A. Noll
PSYCHIATRY Tom Burns
PSYCHOANALYSIS Daniel Pick
PSYCHOLOGY Gillian Butler and
Freda McManus
PSYCHOLOGY OF MUSIC
Elizabeth Hellmuth Margulis
PSYCHOPATHY Essi Viding
PSYCHOTHERAPY Tom Burns and
Eva Burns-Lundgren
PUBLIC ADMINISTRATION
Stella Z. Theodoulou and Ravi K. Roy
PUBLIC HEALTH Virginia Berridge
PURITANISM Francis J. Bremer
THE QUAKERS Pink Dandelion
QUANTUM THEORY
John Polkinghorne
RACISM Ali Rattansi
RADIOACTIVITY Claudio Tuniz

RASTAFARI Ennis B. Edmonds
READING Belinda Jack
THE REAGAN REVOLUTION Gil Troy
REALITY Jan Westerhoff
THE REFORMATION Peter Marshall
RELATIVITY Russell Stannard
RELIGION IN AMERICA Timothy Beal
THE RENAISSANCE Jerry Brotton
RENAISSANCE ART
    Geraldine A. Johnson
REPTILES T. S. Kemp
REVOLUTIONS Jack A. Goldstone
RHETORIC Richard Toye
RISK Baruch Fischhoff and John Kadvany
RITUAL Barry Stephenson
RIVERS Nick Middleton
ROBOTICS Alan Winfield
ROCKS Jan Zalasiewicz
ROMAN BRITAIN Peter Salway
THE ROMAN EMPIRE
    Christopher Kelly
THE ROMAN REPUBLIC
    David M. Gwynn
ROMANTICISM Michael Ferber
ROUSSEAU Robert Wokler
RUSSELL A. C. Grayling
RUSSIAN HISTORY Geoffrey Hosking
RUSSIAN LITERATURE Catriona Kelly
THE RUSSIAN REVOLUTION
    S. A. Smith
SAINTS Simon Yarrow
SAVANNAS Peter A. Furley
SCEPTICISM Duncan Pritchard
SCHIZOPHRENIA Chris Frith and
    Eve Johnstone
SCHOPENHAUER Christopher Janaway
SCIENCE AND RELIGION
    Thomas Dixon
SCIENCE FICTION David Seed
THE SCIENTIFIC REVOLUTION
    Lawrence M. Principe
SCOTLAND Rab Houston
SECULARISM Andrew Copson
SEXUAL SELECTION Marlene Zuk and
    Leigh W. Simmons
SEXUALITY Véronique Mottier
SHAKESPEARE'S COMEDIES
    Bart van Es
SHAKESPEARE'S SONNETS AND
    POEMS Jonathan F. S. Post

SHAKESPEARE'S TRAGEDIES
    Stanley Wells
SIKHISM Eleanor Nesbitt
THE SILK ROAD James A. Millward
SLANG Jonathon Green
SLEEP Steven W. Lockley and
    Russell G. Foster
SOCIAL AND CULTURAL
    ANTHROPOLOGY
    John Monaghan and Peter Just
SOCIAL PSYCHOLOGY Richard J. Crisp
SOCIAL WORK Sally Holland and
    Jonathan Scourfield
SOCIALISM Michael Newman
SOCIOLINGUISTICS John Edwards
SOCIOLOGY Steve Bruce
SOCRATES C. C. W. Taylor
SOUND Mike Goldsmith
SOUTHEAST ASIA James R. Rush
THE SOVIET UNION Stephen Lovell
THE SPANISH CIVIL WAR
    Helen Graham
SPANISH LITERATURE Jo Labanyi
SPINOZA Roger Scruton
SPIRITUALITY Philip Sheldrake
SPORT Mike Cronin
STARS Andrew King
STATISTICS David J. Hand
STEM CELLS Jonathan Slack
STOICISM Brad Inwood
STRUCTURAL ENGINEERING
    David Blockley
STUART BRITAIN John Morrill
SUPERCONDUCTIVITY
    Stephen Blundell
SYMMETRY Ian Stewart
SYNAESTHESIA Julia Simner
SYNTHETIC BIOLOGY Jamie A. Davies
TAXATION Stephen Smith
TEETH Peter S. Ungar
TELESCOPES Geoff Cottrell
TERRORISM Charles Townshend
THEATRE Marvin Carlson
THEOLOGY David F. Ford
THINKING AND REASONING
    Jonathan St B. T. Evans
THOMAS AQUINAS Fergus Kerr
THOUGHT Tim Bayne
TIBETAN BUDDHISM
    Matthew T. Kapstein

TOCQUEVILLE Harvey C. Mansfield
TRAGEDY Adrian Poole
TRANSLATION Matthew Reynolds
THE TREATY OF VERSAILLES
    Michael S. Neiberg
THE TROJAN WAR Eric H. Cline
TRUST Katherine Hawley
THE TUDORS John Guy
TWENTIETH-CENTURY BRITAIN
    Kenneth O. Morgan
TYPOGRAPHY Paul Luna
THE UNITED NATIONS
    Jussi M. Hanhimäki
UNIVERSITIES AND COLLEGES
    David Palfreyman and Paul Temple
THE U.S. CONGRESS Donald A. Ritchie
THE U.S. CONSTITUTION
    David J. Bodenhamer
THE U.S. SUPREME COURT
    Linda Greenhouse
UTILITARIANISM
    Katarzyna de Lazari-Radek
    and Peter Singer

UTOPIANISM Lyman Tower Sargent
VETERINARY SCIENCE James Yeates
THE VIKINGS Julian D. Richards
VIRUSES Dorothy H. Crawford
VOLTAIRE Nicholas Cronk
WAR AND TECHNOLOGY
    Alex Roland
WATER John Finney
WAVES Mike Goldsmith
WEATHER Storm Dunlop
THE WELFARE STATE David Garland
WILLIAM SHAKESPEARE
    Stanley Wells
WITCHCRAFT Malcolm Gaskill
WITTGENSTEIN A. C. Grayling
WORK Stephen Fineman
WORLD MUSIC Philip Bohlman
THE WORLD TRADE
    ORGANIZATION Amrita Narlikar
WORLD WAR II Gerhard L. Weinberg
WRITING AND SCRIPT
    Andrew Robinson
ZIONISM Michael Stanislawski

Available soon:
TOPOLOGY Richard Earl
KOREA Michael J. Seth
TIDES David George Bowers
    and Emyr Martyn Roberts

TRIGONOMETRY
    Glen Van Brummelen
NIELS BOHR J. L. Heilbron

For more information visit our website

www.oup.com/vsi/

Nick Jenkins

ENERGY SYSTEMS

A Very Short Introduction

OXFORD
UNIVERSITY PRESS

Great Clarendon Street, Oxford, OX2 6DP,
United Kingdom

Oxford University Press is a department of the University of Oxford.
It furthers the University's objective of excellence in research, scholarship,
and education by publishing worldwide. Oxford is a registered trade mark of
Oxford University Press in the UK and in certain other countries

Published in the United States of America by Oxford University Press
198 Madison Avenue, New York, NY 10016, United States of America

British Library Cataloguing in Publication Data
Data available

Library of Congress Control Number: 2019946778

ISBN 978-0-19-881392-7

Printed and bound by
CPI Group (UK) Ltd, Croydon, CR0 4YY

Contents

Acknowledgements xv

List of illustrations xvii

Abbreviations and units xix

Useful definitions xxi

Introduction 1

1 Energy systems 5

2 Fossil fuels 19

3 Electricity systems 48

4 Nuclear power 74

5 Renewable energy systems 85

6 Future energy systems 116

Further reading 131

Index 135

List of Illustrations

Acknowledgements

I would like to express my thanks and appreciation to friends and colleagues in the research group at Cardiff University, particularly Prof Jianzhong Wu, Lahiru Jayasuria for Figure 8, and Dr Muditha Abeysekera for Figure 25.

List of illustrations

1 Worldwide energy use (excluding traditional biomass) **11**
From BP Statistical Review of World Energy 2018 (used with permission).

2 Formation of coal (low rank → high rank) **21**

3 Combustion system of a coal fired power station **27**
From Roberts, L. E. J. et al. *Power Generation and the Environment*. Oxford University Press. © Roberts, Liss, and Saunders, 1990. Reproduced with permission of the Licensor through PLSclear.

4 Schematic of an anticline trap forming an oil and gas reservoir **32**

5 Typical life cycle of an oil field **36**

6 Schematic geology of natural gas **42**
U.S. Energy Information Administration and U.S. Geological Survey (Sep 2018).

7 Combined cycle gas turbine **44**
Weedy, B., Cory, B., Jenkins, N., Ekanayake, J. and Strbac, G. (2012). *Electric Power Systems*. 5th ed. John Wiley & Sons. Copyright © 2012, John Wiley and Sons.

8 Simplified schematic of the GB natural gas system **46**
Courtesy of Lahiru Jayasuria.

9 Direct and alternating current **50**

10 Interconnected electricity system **53**
Reproduced with permission from Jenkins, N. and Ekanayake, J. (2017). *Renewable Energy Engineering*. Cambridge University Press.

11 Load on the GB power system **57**
National Grid (used with permission).

12 Merit order, showing operating point (O) **64**

13 Impact of photovoltaic generation on the net electricity demand in California **66**

Redrawn from Cal ISO Fast Facts 'What the duck curve tells us about managing the grid' 2016. Licensed with permission from the California ISO. Any statements, conclusions, summaries or other commentaries expressed herein do not reflect the opinions or endorsement of the California ISO.

14 Schematic diagram of a pressurized-water reactor (PWR) **79**

Weedy, B., Cory, B., Jenkins, N., Ekanayake, J. and Strbac, G. (2012). *Electric Power Systems*. 5th ed. John Wiley & Sons. Copyright © 2012, John Wiley and Sons.

15 Schematic diagram of a boiling-water reactor (BWR) **79**

Weedy, B., Cory, B., Jenkins, N., Ekanayake, J. and Strbac, G. (2012). *Electric Power Systems*. 5th ed. John Wiley & Sons. Copyright © 2012, John Wiley and Sons.

16 Solar irradiance on a horizontal surface in California over five days of January **89**

Weedy, B., Cory, B., Jenkins, N., Ekanayake, J. and Strbac, G. (2012). *Electric Power Systems*. 5th ed. John Wiley & Sons. Copyright © 2012, John Wiley and Sons.

17 Schematic of a poly-crystalline silicon solar cell **91**

18 Parabolic trough solar collector **95**

Reproduced with permission from Jenkins, N. and Ekanayake, J. (2017). *Renewable Energy Engineering*. Cambridge University Press.

19 Average Daily Flow of a river with variable discharge over the year **98**

Reproduced with permission from Jenkins, N. and Ekanayake, J. (2017). *Renewable Energy Engineering*. Cambridge University Press.

20 Schematic diagram of a high head hydro scheme **99**

Reproduced with permission from Jenkins, N. and Ekanayake, J. (2017). *Renewable Energy Engineering*. Cambridge University Press.

21 Largest commercially available wind turbines **101**

22 Wind turbine **102**

23 Output power of a small wind farm **104**

24 Trilemma of energy policy **117**

25 Possible interactions between different energy systems **128**

Courtesy of Muditha Abeysekera.

Abbreviations and units

ac	alternating current
CCGT	Combined Cycle Gas Turbine
CH$_4$	methane
CHP	Combined Heat and Power
CO$_2$	carbon dioxide
CSP	Concentrated Solar Power
dc	direct current
GB	Great Britain; England, Scotland, and Wales, that have interconnected electricity and gas systems
HVdc	high voltage direct current
ICT	Information and Communications Technology
J	Joule, unit of energy equivalent to 1 Watt of power for 1 second
kA	kiloAmps, unit of current equal to 1000 Amps
kV	kiloVolts, unit of voltage equal to 1000 Volts
kWh	kiloWatt-hour, unit of energy equivalent to 1000 W of power for 1 hour
MW	MegaWatt, unit of power equal to 1 million Watts
Nm^3	normal cubic metres. Cubic metres of gas at a temperature of 0 °C and pressure 101.325 kPa
NO$_x$	nitrogen oxides

NTS	GB natural gas National Transmission System
OCGT	Open Cycle Gas Turbine
ppm	parts per million
SO$_2$	sulphur dioxide
W$_e$	electrical power
W$_{th}$	thermal power (heat)

Useful definitions

Capacity factor is the ratio of energy generated over a period (typically a day or year) to that if the generator had operated continuously at full output.

Diversity refers to the reduction in aggregate power demand (or output) from a group of loads (or generators) operating independently. It is the basis on which traditional electricity networks were designed.

Electrical load describes the consumption of electrical power.

Electrical network losses. Network technical losses are the Power (P) converted into heat by current (I) flowing in a circuit of Resistance (R). These losses are proportional to the square of the current $(P = I^2 R)$. There are also some constant technical losses and some non-technical losses.

Insolation (J/m^2 or kWh/m^2) is the energy of the solar resource over a specified time (typically a day).

Irradiance (W/m^2) is the power of the solar resource.

Islanded operation describes generators not connected to a large electricity network. In islanded operation, a generator must create its own voltage and frequency reference while when grid connected these are taken from the large network.

Levelized Costs of Energy (LCOE) are the total costs of the generator over its lifetime divided by the total energy generated.

Marginal cost of generation (MC) is the cost of generating the next unit of electricity and so includes fuel and operation and maintenance costs but excludes the capital costs of constructing the plant.

The *Rule of 72* is a simplified way to estimate how long a quantity will take to double, if it increases at a constant percentage rate. By dividing 72 by the annual percentage rate of increase, an estimate is obtained of how many years it will take for the quantity to double. Thus if energy use increases by a constant 2 per cent/year it will double in approximately $72/2 = 36$ years.

Introduction

Energy systems provide the solid, liquid, and gaseous fuels as well as electricity that we use in buildings, industry, transport, and agriculture. They are essential for modern life but, throughout the world, energy systems are changing at a pace not seen since electricity began to be distributed widely more than 100 years ago. The radical changes in both the supply and use of energy are being driven by the needs of modern society, environmental concerns, and rapid developments in technology.

The population of the world is at present approaching 8 billion and is expected to peak at between 10 and 12 billion sometime this century. Approximately 1 billion people, living mainly in sub-Saharan Africa and rural Asia, have no access to electricity and so are denied simple necessities such as lights, fans, and refrigerators. They are cut off from the Internet and much economic activity unless they find other ways to recharge the batteries of their mobile phones, for example from communal charging points. It is also estimated that around 3 billion people do not have access to clean energy for cooking and rely on open fires and traditional biomass, with damaging consequences for their health and the local eco-systems.

Burning fossil fuels has a severe effect on the environment of the planet at a local level, as manifested in the poor air quality in many cities, and globally is leading to climate change. Although there are some vocal contrarians, the position taken throughout this book is that of most climate scientists and policy makers; that man-made climate change caused by burning carbon based fossil fuels is real and extremely dangerous. To limit global warming to an increase of surface temperature of 1.5 °C above pre-industrial levels requires net zero emission of CO_2 by around 2050. While the transition to low carbon energy systems (particularly electricity generation) is already occurring, it will be a huge challenge to move away from fossil fuels. More optimistically, there are many environmental and health benefits as well as commercial opportunities in making this transition.

Electrical energy has very low environmental impact at its point of use and is increasingly important as a means to distribute energy. It powers the knowledge economy and electric cars are becoming common in countries as different as Norway and Sri Lanka. Partly because of its lower environmental impact and partly because of convenience and cost, natural gas has been displacing coal as the fuel for some new electricity generation. This change has been accelerated by the rapid increase in the production of gas from shale and other unconventional sources in the USA. Low carbon energy sources particularly solar and wind energy have recently shown such dramatic reductions in cost that, under favourable conditions, they are now competing with electricity generated from fossil fuels without government subsidy.

The increase in electricity generated from renewables is changing the way power grids are operated. The proliferation of generators powered by renewable energy is leading to a radical reappraisal of the role of traditional integrated electricity utilities that have built their businesses based on interconnected transmission networks supplied by large central (often fossil fuelled) generators. As the cost of storing electricity in batteries falls, local generation from renewable energy sources will increase and the distinction between the producers and consumers of electrical energy will become more and more blurred. The revolution in information and communication technology and the ease with which large volumes of data can now be handled is supporting the development of the Smart Grid.

In the 1950s it was thought that electricity generated from nuclear energy would be universally attractive and would be 'too cheap to meter'. Nuclear generation remains controversial due to its high costs, concerns over safety, and the slow progress that is being made in establishing satisfactory facilities for the disposal of radioactive waste. Environmentalists are divided over whether the advantages that nuclear power offers of constant electricity generation with minimal emissions of greenhouse gases outweigh the risks and environmental impacts.

All options for energy supply entail compromise and require hard choices to be made. One desirable course of action is for the richer inhabitants of the world to reduce their consumption of energy. Although not dealt with in this short book in the detail that the topics deserve, energy demand reduction and energy conservation should not be neglected as we consider energy supply systems.

In such a rapidly evolving energy landscape the need for sound understanding to allow informed debate is paramount. However, much of the current discourse on energy systems is not clearly expressed or easily followed. Common examples of the confusion that is found in many otherwise reputable sources of information are: not recognizing the difference between energy and power, confusing electrical and thermal energy, and data being quoted in units that do not exist such as kW/hour. Particular care is needed when comparing costs of energy from different energy systems as these depend heavily on assumptions and how the calculation is made.

The ambition of this book is to assist students and anyone with an interest in energy questions to acquire a basic understanding and vocabulary that will encourage their participation in the crucially important debate of how environmentally acceptable and affordable energy can be provided for all. Thus the book starts with a review of basic ideas and well-established energy systems before moving on to consider how a future energy system might be configured. In this book our definition of an energy system includes the primary fuel, conversion system, and means of moving energy to its point of use.

Chapter 1
Energy systems

A reliable and affordable supply of energy is essential for modern life and allows us to cook food and distribute water, as well as to heat and light our homes. Energy is needed in agriculture for cultivation and the manufacture of fertilizer. It is used for transport and to process materials into manufactured goods while communication and computer systems consume increasing quantities of electrical energy. Thus it is impossible to conceive of a modern society without reliable and effective systems to provide energy when and where it is needed.

Until the 18th century, energy was supplied mainly from animal power or by burning biological material such as wood. Coal then started to be exploited widely as a source of heat and to fuel the steam engines of the industrial revolution. Animal products (particularly whale oil) were used for lighting until mineral oil was found and exploited in commercial quantities in the mid-19th century. In the 19th and early 20th centuries, the convenience of transporting energy through pipes led to gas works being established in many towns and cities to manufacture gas from coal; networks of gas pipes were constructed and many of the pipes (sometimes relined) are still used to transport natural gas. Electricity became an important way to distribute energy following the construction of the first electricity generating power stations and local distribution networks in the 1880s, some 140 years ago.

Modern energy systems are a comparatively recent development and they continue to evolve in response to the demand for energy and to the changing sources of supply. Increasingly environmental considerations, as well as costs, determine which energy sources are used and how the energy transmission and distribution networks are constructed and operated.

Energy and power

Energy is defined as the ability to do work such as move a vehicle, or heat and light a room. Common forms of energy include: heat, light, motion, electricity, chemical, and nuclear energy. Energy can neither be created nor destroyed only changed in form; for example as electrical energy operates a computer it is converted into an equivalent quantity of heat energy. When energy is transformed from one form to another some of the energy is lost into the atmosphere as low temperature heat. For example, an electric motor consuming 10 kWh of electrical energy will typically produce 9 kWh of mechanical energy and pass 1 kWh of heat energy into the surrounding air.

Some forms of energy are more flexible and hence more valuable than others. Electricity can be used to operate a telecommunications system or boil water, while gas or coal can be burnt to give heat but cannot power a computer. The usefulness of thermal energy depends on its temperature, with high temperatures being required for efficient conversion into mechanical energy. An electrical generator powered by a steam turbine will typically only turn about 40 per cent of the energy in the fuel into electricity and reject 60 per cent as heat into the atmosphere. Most of this difference is caused by limits to the temperature of the steam that can be used in the boiler and turbine. Thus all forms of energy are not interchangeable and although the overall amount of energy is conserved it is important to define clearly the type of energy that is being discussed.

The main sources of energy in the world are the renewable energy of the sun and non-renewable fossil fuels. Energy from the sun strikes the earth as solar radiation and directly provides heat and light. It is also concentrated through various mechanisms such as photosynthesis to create biomass, the creation of winds in the atmosphere, and the hydrological cycle that results in rain. As the life of the sun is effectively infinite, these energy sources are termed renewable. Fossil fuels (coal, oil, and natural gas) were formed from decaying biomass that was created over many millions of years originally by photosynthesis using the energy from the sun. However they are now being exploited so much more rapidly than the rate at which they were created that they are non-renewable. When atoms of uranium are split in the process of nuclear fission a very large quantity of heat is released that can be used to generate electrical energy. In its simplest implementation, nuclear fission requires a source of uranium, which is a finite resource.

Discussions of energy and energy systems are often confused by the loose use of terminology and units. It is particularly important to be clear over the difference between energy and power. Power is the rate at which energy is transformed and so power can be generated from all the forms of energy. Thus power can be electrical, mechanical, thermal, or hydraulic. Electricity is only one of the many forms of energy or power, if a very convenient one. In many temperate countries only about a third of the energy used nationally is electrical energy; the remainder is obtained from burning fossil fuels directly for heating or from oil for transport. However, in general conversation the term 'power' is often used loosely to refer to electrical power and even electrical energy. For an informed conversation it is essential to distinguish clearly between energy and electrical energy, and energy and power.

There is a (perhaps apocryphal) story of a government minister negotiating an international agreement of how much energy should come from renewable sources. He agreed that 15 per cent of national energy supply should come from renewable sources while thinking he was agreeing that 15 per cent of electrical energy would come from renewable sources. As the total energy use of the country was three times the amount of electricity used, the minister had unwittingly committed his government to a target three times higher than he had thought!

In this short book we will try to be precise in distinguishing between energy, power, and electrical energy and electrical power. Box 1 is a simple example illustrating the difference between energy and power while Box 2 shows the usual units that are used in discussions of energy.

Box 1 Illustration of the difference between energy and power

A domestic kettle contains 1 litre of water. To heat the water from room temperature of 20 °C to its boiling point of 100 °C takes around 336 kJ of energy. If an electric kettle has a power rating of 2 kW it transforms electrical energy into heat at a rate of 2 kJ/s or 2 kW. Thus the kettle will take 168 seconds (336/2) or 2.8 minutes (168/60) to warm the water. In this simple example (which neglects heat lost to the atmosphere) the energy required is 336 kJ and the power used in the kettle is 2 kW for 2.8 minutes. A lower power kettle of 1 kW rating would still provide the 336 kJ of heat energy required but would take 5.6 minutes (2×2.8) to heat the litre of water. The energy of 336 kJ (Ws) can also be described as 93 Wh (0.093 kWh) by dividing the energy (in joules) by the number of seconds in an hour, i.e. 3600.

Box 2 Units used to quantify energy

A **joule (J)** is the SI (Système International) unit of energy. It is equal to 1 watt of power for 1 second.

A **kilowatt-hour (kWh)** is the usual unit used for electrical energy produced or consumed. It is 1000 Wh and so is equal to 1000 W of power for 1 hour. 1 kWh is equal to 3.6×10^6 J.

However larger generators produce many thousands of kWh described in MWh, GWh, and TWh. The relationships between these are:

1 MegaWatt-hour, MWh $= 10^6$ Wh $= 1000$ kWh

1 GigaWatt-hour, GWh $= 10^9$ Wh $= 1000$ MWh

1 TerraWatt-hour, TWh $= 10^{12}$ Wh $= 1000$ GWh

One tonne of oil equivalent (toe) is the energy released by burning one tonne of oil. When discussing various energy systems, it is convenient to use a single unit and so energy from different fuels is converted into the energy produced by burning one tonne (1000 kg) of oil. The heat content of one tonne of crude oil depends on its origin but is generally taken as 41.87 GJ (10^9 J).

The following conversions are usually used:

The heat value of 1 tonne of coal $= 29.3$ GJ $= 0.7$ tonnes of oil equivalent

1 MWh of electrical energy $= 3600 \times 10^6$ J $= 3.6$ GJ $= 0.086$ tonnes of oil equivalent

1 kg of oil equivalent is equal to 11.63 kWh.

In these conversions of electricity generated into tonnes of oil equivalent it is assumed that the efficiency of the thermal generating plant is 100 per cent. If the efficiency of the plant is

(*continued*)

taken as 38 per cent, a realistic figure for modern steam turbine thermal generators, then 1 kg of oil equivalent produces 4.42 kWh of electrical energy (11.63 kWh × 0.38).

A Quad (Quadrillion BTU) is a unit used in North America to quantify various forms of energy. It is defined as 10^{15} BTU (British Thermal Units). Annual energy consumption in N. America is presently about 120 Quads and worldwide consumption is around 520 Quads. 1 Quad is equal to approximately 25.2 Mtoe.

Worldwide use of energy

Figure 1 shows the worldwide use of energy from 1992 to 2017. In 2017 around 13.5 Gtoe of primary energy was used, mainly from fossil fuels (coal, oil, and natural gas). In addition to the traded energy shown in Figure 1, traditional biomass is used for heating and cooking in the rural areas of many developing countries. Over the last fifty years worldwide annual energy use has increased at an average annual rate of 2½ per cent/year. The annual rate of increase of energy consumption worldwide has dropped slightly in recent years to around 2 per cent/year depending on the state of the world economy. A 2 per cent annual increase leads to a doubling in energy use every thirty-six years (Rule of 72).

Table 1 shows the sources of primary energy consumed in 2017. The 15 per cent of energy not from fossil fuels was supplied by hydro (7 per cent), nuclear (4 per cent), and renewables excluding hydro (4 per cent), almost all as electrical energy. The table also shows how energy is used for generating electricity. This is still dominated by coal (38 per cent) although with an increasing fraction from natural gas. Renewables (excluding hydro) provided 4 per cent of worldwide primary energy but 8 per cent of electrical energy. In that year solar and wind energy systems were installed

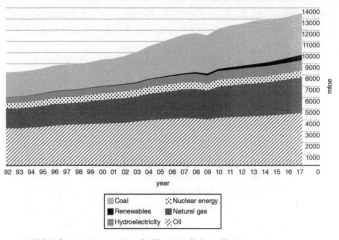

1. Worldwide energy use (excluding traditional biomass). Million tonnes of oil equivalent—mtoe.

Table 1. Sources of primary energy consumed and electricity generated worldwide

Source	% of energy consumed	% of electricity generated
Oil	34	4
Natural gas	23	23
Coal	28	38
Nuclear energy	4	10
Hydro	7	16
Renewables ex hydro	4	8

BP Statistical Review of World Energy 2018 (used with permission).

at a faster rate than other forms of electricity generation plant but from a low base.

The fossil fuels that we currently use were laid down over millions of years and have been exploited in significant quantities for less than three centuries. The two obvious difficulties in continuing to use fossil fuels at the rates shown in Figure 1 are the availability of the resources and the environmental impact of burning fossil fuel.

The availability of a fossil fuel can be described by its Reserves/Production Ratio. This is defined as the ratio of the proved reserves of the fuel that can be extracted using known production techniques and at reasonable cost, divided by the annual production. For oil and gas, the worldwide Reserves/Production Ratio of each fuel is at present around fifty years and for coal around 100 years. The Reserves/Production Ratios have to be used with care as although annual production can be determined fairly precisely, the extent of the proved reserves that are left to be exploited relies on a judgement of the remaining fossil fuel that can be extracted under current economic and operating conditions. This changes with time as reserves are discovered and new extraction methods developed.

The Reserves/Production Ratios do not predict that oil and gas will run out suddenly after fifty years (and coal after 100 years) but rather warn that reserves are finite and the cost of extraction will rise and prices increase as those reserves that can be extracted easily and cheaply become exhausted. A particular problem in determining the reserves of oil is that several major oil producing countries rely heavily on oil revenues to support their economy and the living standards of their population. Thus the size of their oil reserves is politically sensitive and publicly available data may not be reliable. However, it is generally agreed that the scarcity of easily extracted resources of conventional oil and gas will cause production to be reduced dramatically by the end of this century.

Although there are around 70,000 oil fields worldwide, 50 per cent of production comes from 100 very large fields, most of which were discovered before 1970, are well understood, and now past their peak production. More recent new discoveries have tended to be in areas of the world where production is difficult, expensive, and with considerable potential environmental impact. The typical

life cycle of an oil well has been that output rises to a maximum after it is developed and then declines over 30–50 years.
The current decline in production from the UK continental shelf is a clear example of this behaviour of gas and oil fields. By assuming this behaviour and examining the rates of discovery of new reserves of oil, it can be concluded that the production of oil and natural gas from conventional sources is now past its peak in many countries.

Some commentators dispute the idea of any impending shortage of oil and gas, and consequent sustained high prices. They argue that unconventional reserves, such as shale oil and gas or tar sands, together with the new production techniques of directional drilling and hydraulic fracturing (fracking), will lead to a surplus of supply of fossil fuel over demand for the foreseeable future. The evidence to support this view is mainly from the USA where the extraction of shale and tight oil and gas has increased considerably in recent years due to developments in well technology, in addition to the particular landownership structure of that country that gives the subsurface mineral rights to the owner of the land's surface. World supplies of coal appear to be adequate to meet demand for the foreseeable future.

Environmental consequences of burning fossil fuels

A more pressing objection to the continued use of fossil fuels is the impact on the environment. The consequences of burning fossil fuel can be considered as the local effect on air quality (particularly in urban areas), the regional effects of acid rain, and the global effects of climate change.

The air quality of many cities is now seriously impaired by emissions from the internal combustion engines of vehicles, fuelled either by petrol (gasoline) or diesel. The main emissions of vehicle engines are nitrogen oxides, carbon monoxide, volatile organic compounds, and particulates. Fossil fuelled heating systems of buildings and industrial plants produce similar emissions and local impacts. Bad air quality in a number of large cities is a significant cause of poor human health, particularly where atmospheric conditions result in concentration of the pollutants.

Burning coal in power stations produces sulphur dioxide (SO_2) and other pollutants including nitrogen oxides (NO_x) that when emitted to the atmosphere lead to acid rain and considerable environmental damage, particularly to buildings, lakes, and forests. Acid rain, created by emissions from coal power plants with high chimneys, has a regional effect with, for example, British power stations causing environmental damage in Germany and Scandinavia as the prevailing winds blow from the west. Acid rain has been recognized as an important environmental problem in parts of the USA and regulations limit the sulphur content of coal that can be burned in power stations. In Europe large coal fired power stations must be fitted with equipment to limit the emissions of SO_2 and NO_x in order to continue operating. Suitable equipment that captures these flue gases is available but adds cost and takes a significant amount of electrical energy to operate, thus reducing the efficiency of the generating unit.

There is a clear scientific consensus that the earth's climate is being changed by human activity through the greenhouse effect, as heat from the earth that would otherwise be radiated into space is trapped by increased concentrations of gases in the atmosphere. The main greenhouse gases are: carbon dioxide (CO_2), methane (CH_4), nitrogen oxides (NO_x), and fluorocarbons. Water vapour also plays an important role in the greenhouse effect. The greenhouse effect is a most complicated phenomenon and the different gases and aerosols in the atmosphere can either increase or lower the earth's temperature. However, it can be thought of simply as the effect of gases in the atmosphere trapping the long wavelength radiation that is emitted from the earth's surface.

The sun is a high temperature source of energy with an effective temperature at its outer surface of around 5800 °C. It emits short wavelength (high frequency) radiation that passes through the earth's atmosphere largely unchanged. This radiation strikes the earth and warms it; the earth then re-radiates long wavelength (low frequency) radiation from its lower surface temperature of 15–20 °C. Gases in the atmosphere trap the lower frequency (longer wavelength) radiation emitted from the earth so warming the atmosphere and in turn the earth.

The temperature of the earth depends on the balance between the incoming energy from the sun and how much is re-radiated from the earth's surface. The concentration of existing greenhouse gases in the earth's atmosphere causes the temperature of the earth to be maintained at a level suitable for life; without it the earth would be colder by some 35 °C. By increasing the concentration of greenhouse gases in the atmosphere, more of the low frequency radiation is trapped and so the temperature of the earth increases. An increase in the average temperature of the earth's surface has significant implications for agriculture and wildlife. Moreover the consequent effects, such as an increase in sea level due to melting of the ice in Greenland and the Antarctic, and the increase in frequency and intensity of extreme weather events, are potentially even more serious. Greenhouse gases disperse throughout the earth's atmosphere and so their effect is global.

Carbon dioxide is an inevitable product of burning fossil fuels and once emitted it remains in the atmosphere for up to several hundred years. There is some uncertainty of the precise length of time that CO_2 stays in the atmosphere and how it is absorbed in the oceans. CO_2 is one of the most important greenhouse gases and its increased concentration in the atmosphere is correlated with warming of the earth and extreme weather events. Hence many countries have policies to reduce emissions of CO_2, particularly from electricity generation. These actions are coordinated through the United Nations Framework Convention on Climate Change (UNFCCC). Discernible progress is being made in many countries to decarbonize electricity supply but it is proving more difficult to find ways to reduce CO_2 emissions from heating, industry, and from heavy transport.

Limiting energy use

The increase in energy use over the twenty-five years shown in Figure 1 is due partly to the increase in the world's population over this period from around 5.2 billion to 7.2 billion but also to many countries increasing their per capita energy consumption as the wealth of their population increases and living standards improve. It is anticipated that the per capita energy use in many large rapidly developing countries will increase from the current world average of 1000 kgoe/year towards the average consumption of more developed countries of 2000–3000 kgoe/year.

In response to both the cost and environmental impact of supplying energy, many countries have policies and programmes to limit their energy use either by using energy more efficiently to achieve the same outcomes or by encouraging their citizens to change their behaviour and use less energy. These measures have had only limited success. Individuals and organizations are often more concerned with convenience and immediate expenditure than with benefits that accrue in the future. Also, commercial incentives may not align; for example in rented buildings it is often the case that expenditure to improve energy efficiency falls on the landlord while the tenant stands to benefit from the future savings. Measures to reduce energy use that rely on individuals changing their behaviour have been even more difficult to implement.

Energy consumption is stabilizing in some developed economies but this is mainly because industrial production has moved to lower cost countries, so merely shifting the energy use and emissions. Thus although the global annual rate of increase of world energy use may drop below 2 per cent in the future, it is likely that increases will be greater in those countries that are industrializing.

Density and storage of energy

The fossil fuels of coal, oil, and gas all have the important characteristics of being dense forms of energy and can be stored for months or even years in simple containers without degradation or excessive losses. Diesel oil has an energy density of around 13 kWh/kg and can be kept in tanks and transported by tanker or through pipelines. Bituminous coal has an energy density of 7 kWh/kg and is easily stored in heaps outdoors, while natural gas can be stored in pressurized pipelines or tanks at high pressure with an energy density of 14 kWh/kg. Nuclear energy is an extremely concentrated form of energy.

The high energy density of fossil fuels and the low cost of storage are in contrast to renewable energy. Most renewable energy is derived from sunlight, which is a diffuse energy source with a maximum irradiance power at the earth's surface of only around 1 kW/m^2 and provides energy at a maximum rate of 6–8 kWh/m^2. day. Although some forms of renewable energy (e.g. wind, biomass, and hydro) concentrate the solar energy, the energy density of almost all renewable sources is much lower than that of fossil fuels.

Many forms of renewables are converted into electrical energy. However there is as yet no practical way to store large quantities of electrical energy cheaply and electrical systems have to be operated by balancing the power generated and the demand on a second-by-second basis.

There are certain technologies and locations where electrical energy can be stored, for example in chemical form in batteries or through water raised against gravity in pumped storage schemes, but the use of electricity storage is small when compared to global energy demand. A conventional Lithium-Ion battery has an energy density of around 0.25 kWh/kg and a rate of self-discharge of 1–2 per cent per month. Although dropping in price, batteries remain expensive and power electronic converters are needed to convert the direct current of the battery to the alternating current of the grid network. The search for technologies to store electricity in large quantities and cost-effectively remains a work in progress.

Chapter 2
Fossil fuels

For the last 300 years the world has relied on fossil fuels for its energy, initially using coal, then oil, and more recently natural gas. Figure 1 showed how fossil fuels continue to supply most of the world's energy. The widespread availability, high energy density, ease of transport and storage, and the generally low cost of fossil fuels has facilitated the development of modern society with all its benefits. The early factories of the industrial revolution had to be located where waterpower was available, but the convenience of coal together with the invention of the steam engine allowed factories to be sited wherever a workforce and raw materials were available. Coal and steam engines stimulated the development of the railways while more recently fuels derived from oil have been used to power the great majority of vehicles. Most of the electrical energy used in the world at present is generated from fossil fuels, particularly from coal, while kerosene derived from oil is essential for powering aircraft engines.

However, the extraction and combustion of fossil fuels has severe impacts on the environment. Some effects are local and can be managed while for others there are as yet no cost-effective solutions. It is accepted by most policy makers that continuing to burn carbon based fossil fuels cannot be sustained and we need to move quickly towards energy systems that emit greatly reduced amounts of CO_2.

History teaches us that bringing any fundamentally new energy technology to market takes more than 40–50 years from the demonstration of the initial idea to widespread commercial acceptance. For example, modern photovoltaic cells were demonstrated in the 1950s but this technology is only now making a major impact on electricity systems. Gas turbines were developed for aircraft in the 1940s but only became important for bulk electricity generation after 1980. Hoping for an undiscovered fuel or technology that will act as a silver bullet to solve our energy needs is unrealistic.

However, the speed of transition between different established energy systems is more rapid (often around 10–20 years) when technology developments, commercial conditions, and government policies align. This can be seen in the rapid switch from manufactured town gas to natural gas, the speed with which rural areas were electrified after the Second World War, and the recent move away from coal mining in many European countries. Most major vehicle manufacturers are developing and beginning to market electric cars, whose use will increase electricity demand while reducing the consumption of oil products.

Coal as a source of energy for electricity generation is particularly vulnerable to changes in policy caused by concern over climate change as it emits around twice as much carbon dioxide for each unit of electricity generated as natural gas. New coal fired power stations are expensive to build and need to generate for a large number of hours each year to spread their fixed costs. The increasing fraction of electricity being generated by renewables with very low operating costs is reducing the hours each year that coal power stations are called on to run and so they are vulnerable to becoming 'stranded assets', unable to recover their construction costs. The transition to a low carbon energy system favours flexible gas turbines with low construction but higher operating costs. Although fossil fuels will continue to play an important role in energy supply for some years there is no certainty that each fuel will retain its present position in the energy mix in the longer term or that the established energy companies and their business models will survive in their current form.

Coal

Coal currently supplies around 28 per cent of the world's traded energy but in recent years its production has remained broadly constant. Historically it was widely used for heating and cooking. As a boy in England my daily chore was to fill coal buckets for an open fire and cooking stove and I was very happy when gas fired central heating was installed in the house. Coal is still used to heat buildings in some countries but now its main use in the West is for generating electricity. Overall 38 per cent of the world's electricity is generated from coal and until 1990 it was the dominant fuel for electricity generation in the UK. Its use however is incompatible with a decarbonized electricity system and is now being phased out.

Coal is a heterogeneous, combustible rock that was formed up to 300 million years ago from mainly organic material. Decaying plant matter from primeval swamps created moist peat bogs that were compressed as successive layers of silt and sand covered the biological material in a process of sedimentation. The heat and pressure acting on the submerged decaying organic matter caused the removal of volatile material in an anaerobic process leaving mainly carbon. The pressure and heat over time in turn formed lignite, sub-bituminous, and bituminous coal and anthracite (Figure 2).

peat ➔ lignite ➔ sub-bituminous ➔ bituminous ➔ anthracite

2. **Formation of coal (low rank → high rank).**

21

Coal is classified by 'rank' which indicates the age of the deposit, the fraction of carbon, and the heating value. The oldest and highest rank coal is hard anthracite that consists mainly of carbon. Bituminous coal contains much carbon but still retains some volatile matter. The least mature material, formed around 65 million years ago, is softer lignite or brown coal. Lignite is normally found near the surface in thick layers, or seams, while the hard coals such as anthracite are usually found deeper in thinner seams. It is believed that low sulphur coal was formed in freshwater swamps while the action of bacteria in seawater led to the creation of higher sulphur coal.

Table 2 shows the typical composition of various ranks of coal although there is a large variation depending on the source of the coal, and even significant differences within seams. The heating value is determined mainly by the carbon and hydrogen fractions but also by the moisture content. When coal is burnt, undesirable sulphur oxides are formed as well as nitrous oxides both from nitrogen within the coal and from the air used in combustion. Some mineral matter was mixed with the organic material as the coal was formed and this creates ash when burned. Ash causes fouling of boilers, requires careful disposal, and leads to particulate emissions in the exhaust gas.

Lignite and even peat are used to generate electricity in some countries but they have a relatively low heating value and high moisture content, so large volumes are required. These are expensive to transport and handle, and are usually burned near the mine, often in relatively small capacity power plants. Most large coal fired power stations burn sub-bituminous or bituminous hard coal with low sulphur content, limiting emissions of sulphur oxides.

There are two main approaches to the extraction of coal, depending on the depth of the seam; these are surface or opencast, and underground or deep mining. Coal near the surface, typically less

Table 2. Typical composition of coal (% by weight) and heating value

Coal type	Carbon %	Hydrogen %	Oxygen %	Sulphur %	Ash %	Heating value MJ/kg
Anthracite	90	3	2	1	3	24+
Bituminous	74	6	13	2	5	17–24
Lignite	57	6	32	2	4	10–17

than ~100 m below ground level, is mined by removing the surface soil and rock, known as the overburden, breaking up the seam with excavators, and then using large trucks to extract the coal. Surface mining allows up to 90 per cent of the coal to be recovered and is less labour-intensive and cheaper than deep mining.

Underground or deep mining is usually undertaken by sinking vertical shafts and then driving horizontal tunnels to access the hard coal, which occurs in seams typically up to 3 m thick. Using the traditional approach of room and pillar mining, coal is extracted creating rooms in the seam leaving vertical pillars of coal to support the roof. In its simplest form, room and pillar mining extracts only about 70 per cent of the coal. Deep mining techniques have developed over the years and modern mines can now be highly automated. In one technique, known as longwall mining, several hundred metres of a seam are exposed to create a long coalface and then the coal is cut from the seam using automatic cutters and extracted on conveyer belts. The roof is supported on hydraulic jacks and allowed to collapse as the coal is extracted. However, less automated, more labour-intensive methods continue to be used for underground mining in some countries, particularly in smaller mines. Drift mines extract coal from inclined seams that start near the surface.

The extraction of coal by either surface or deep mining is hazardous and can have serious environmental and health impacts. Surface mining involves the removal of very large volumes of overburden to access the coal, the use of heavy machinery, and often explosives. Moving such large quantities of material inevitably results in the creation of dust and noise while causing long-term damage to topsoil, local ecology, and surface and ground water. Mountaintop removal mining that is practised in parts of the USA is an extreme example of surface mining in which entire mountaintops are removed to expose coal seams. The overburden is disposed of in adjacent valleys and the landscape and environment reinstated once the coal is extracted.

Surface mining has the potential for significant environmental impact, but is less hazardous and less damaging to the health of miners than deep mining.

Historically deep mining of coal has been one of the most hazardous of all occupations both because of accidents and the cumulative impact of the miners' exposure to dust. Countries with poorly developed or unenforced health and safety regulations for mining continue to experience high rates of accidents and poor health of miners. Deep mining has significant environmental impacts through the creation of spoil heaps and subsidence. For every tonne of coal extracted from a deep mine, half a tonne of spoil is created and this has to be stored in spoil heaps that are subsequently stabilized and landscaped. The large volume of material removed by deep mining leads to surface subsidence and damage to buildings. Subsidence typically ranges from a few centimetres to 0.5 m and can occur over an area much wider than the mine workings. Deep mining can also lead to the release of methane and water pollution. Abandoned coal mines can cause significant pollution of water resources both from run-off from spoil heaps and from the water trapped in old mine workings.

Once mined, coal is cleaned in a processing plant to remove rock, ash, and other unwanted material, and crushed into a uniform size to aid handling. Coal processing may be through dry or wet processes, with wet coal processing allowing easier control of dust but creating large volumes of slurry requiring careful disposal. It is common to blend coal from different mines or seams to provide the quality required for particular applications, for example to limit ash or sulphur content.

Coal plays an important role in the manufacture of steel from iron ore and in many countries, for example China, India, and the USA, it continues to be widely used for the generation of electricity. Over recent years India and China have made great progress in giving their populations access to electricity but they rely heavily on coal fired generation.

Figure 3 shows how coal is burnt in a boiler of the type used to make steam for a turbine-generator unit. Hard coal is transported to the power station, often by trains or barges, crushed, and then ground into a fine dust in mills, to improve combustion and ease of handling. The pulverized coal is blown into the boiler together with primary air and burnt. Secondary air from forced draught fans is introduced to create a temperature in the furnace of around 1700 °C. The exhaust gases are dragged through the boiler by induced draught fans and passed though electrostatic precipitators to capture the fly ash. The exhaust gases are vented to the atmosphere through a high stack (chimney) to disperse them.

De-ionized water is forced into the boiler by powerful feed pumps and heated in a series of evaporator tubes around the edges of the furnace. The dry steam is separated and passed to a super-heater stage for further heating. The steam is then used to drive a multi-stage turbine before being returned to the boiler for reheating between turbine stages.

There are two main types of steam generator that burn pulverized coal. These are distinguished by the steam conditions and are drum type or once-through boilers.

The older drum type boilers typically produce steam at 568 °C and 165 bar pressure. A typical net efficiency of the entire electricity generating unit (measured as the electrical energy generated compared to the energy in the coal) would be 30–35 per cent.

A once-through boiler can use steam at higher temperature and pressure, and so the turbine-generator unit operates more efficiently. Modern super-critical, pulverized coal, once-through boilers create steam at 600 °C and 300 bar pressure. These higher

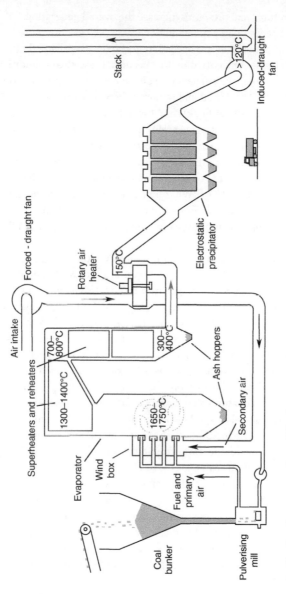

3. Combustion system of a coal fired power station.

temperatures and pressures increase the overall generating unit efficiency up to a maximum of 46–48 per cent but more specialized materials are needed in the boiler to withstand the higher stresses and corrosion that occur in these steam conditions. This higher thermal efficiency results in a 35 per cent reduction in the amount of coal used and importantly reduces emission of CO_2 from around 1000 gCO_2/kWh in the older drum type boilers to about 700 gCO_2/kWh of electricity generated. Although reducing the CO_2 created for each unit of electrical energy is clearly desirable, the carbon intensity of coal fired generation remains much higher than that of generation using natural gas, which is around 400–500 gCO_2/kWh. An interim target for a transition to a decarbonized electricity system compatible with policy ambitions to limit climate change would be 50–100 gCO_2/kWh.

In addition to containing 12–14 per cent CO_2, the flue gas of a pulverized coal boiler typically contains around 1500 ppm of SO_2 and some 500 ppm of nitrogen oxides (NO_x), depending on the chemical composition of the coal and the boiler firing conditions. These emissions can have severe environmental and health impacts and are limited by regulation in most countries. SO_2 is a product of the combustion of sulphur held within the coal, either organically bound or as pyrite (FeS_2). Oxides of nitrogen are formed by the combustion process (creating fuel NO_x) but also by the oxidation of atmospheric nitrogen at high temperatures (creating thermal NO_x). The power station stack (chimney), which may be up to 200 m high, creates an exhaust plume to disperse the pollutants away from heavily populated areas. In dry weather this plume can travel several hundred kilometres before the pollutants are absorbed by land or water surfaces in a process known as dry deposition. When the air is moist water droplets absorb the gases and wet deposition results in increased acidity of the rainfall downwind of the power station. Acid rain has led to serious environmental consequences particularly to lakes and streams and their populations of fish in both Europe and the USA.

Flue Gas Desulphurization equipment (FGD) is used to reduce SO_2 emissions if low sulphur coal is not available. In one common approach the flue gases are passed through a slurry of limestone ($CaCO_3$) in water to form gypsum ($CaSO_4$). The process removes 70–90 per cent of the sulphur dioxide but requires large amounts of limestone and creates similar volumes of wet waste product. The capital and running costs of limestone/gypsum FGD plant increase the cost of electricity generation by around 10 per cent. In an alternative approach, often used with lower sulphur coal, a slurry of lime is sprayed into the flue gases and the latent heat produces a dry waste product. Nitrogen oxides from coal fired power stations can be controlled through a number of approaches, including low NO_X combustion, selective catalytic reduction, and selective non-catalytic reduction. These can be used individually or in combination.

The flue gases of a pulverized fuel boiler contain particulate matter, mainly as dust from the coal ash. This dust is removed by using either bag filters or electrostatic precipitators in the flue gas stream. These consist of a series of electrodes that create an electric field, which attracts dust and small particles to the electrode plates. The plates are struck periodically to loosen the dust that then falls into a hopper for disposal. In the slow flue gas stream from a pulverized fuel furnace an electrostatic precipitator removes more than 99.5 per cent of the dust.

Emissions of both SO_2 and NO_X are both reduced in fluidized bed combustors, which are increasingly being considered for electricity generation. These have a bed of inert sand or limestone that is kept suspended in a turbulent state by a supply of pressurized air from fans. The fuel is burnt in and above the bed, which is typically kept at 800–1050 °C. There are two common types of fluidized bed combustors; the bubbling fluidized bed and the circulating fluidized bed. A bubbling fluidized bed has a zone of suspended inert particles that is supported by the stream of air. Above the bed is a turbulent combustion zone into which the fuel is introduced. In a circulating fluidized bed, the air velocity is higher and some particles of both the inert bed and the burnt fuel flow upward with the gas stream, out of the combustor, and are separated in an external cyclone and returned to the combustion zone. The turbulent conditions of a fluidized bed combustor lead to intimate mixing of the products of combustion and the inert bed so providing good heat transfer. Emissions of NO_x are limited by the lower temperatures compared to a pulverized coal furnace while SO_2 is reduced by the intimate mixing of limestone in the combustion region. A further advantage of fluidized bed combustors is that they are easier to co-fire with other fuels such as biomass.

Burning coal (i.e. oxidizing carbon) inevitably creates large quantities of CO_2 and attempts to control the CO_2 emissions from burning coal are still in the research and demonstration phase. Carbon Capture and Storage (CCS) is the process of capturing CO_2, transporting it to a geological disposal site, and storing it there for a long period. Several approaches are being investigated but the most mature technology for a coal fired power station is to use an amine scrubbing process to remove CO_2 from the flue gas after combustion and transport it via pipeline to an underground or undersea storage site. An alternative, pre-combustion technology uses a gasifier fed with coal and oxygen to manufacture hydrogen and CO_2. The hydrogen produces only water when it is burnt and the concentrated CO_2 is sent for storage. CCS technology has been demonstrated at full scale at one site in Canada but its cost-effectiveness is unclear and there are continuing concerns over the CO_2 leaking from the geological storage sites, and public acceptability. An alternative that is being investigated is carbon capture and utilization (CCU) where the CO_2 is used for chemical manufacture or materials processing.

Oil

Oil currently provides around 34 per cent of global energy and its consumption has been increasing in recent years at around

1 per cent per year. However this rate of increase is not spread uniformly across the world and the annual rate of growth of consumption of oil in both India and China currently exceeds 5 per cent, indicating a doubling of demand every fifteen years. More than half the world's oil is used for vehicle engines. Heavy oil can be burnt to generate electricity in a steam-raising boiler and more refined oil is often used as a liquid fuel for gas turbines. The term gas turbine refers to the thermodynamic cycle not the fuel and so a gas turbine can be powered from gas, kerosene, or even crude oil. However, oil products are only used to generate 4 per cent of the world's electricity and then mainly in oil rich countries.

It is generally accepted that oil and natural gas were formed millions of years ago by decomposing marine plants and micro-organisms such as algae and plankton in a process of anaerobic geological sedimentation similar to the creation of coal. Oil and gas deposits are not found in well-defined seams but are distributed throughout the porous rock that was formed from compressed mud, sand, and silt. The pressures and temperatures experienced by the decomposing organic material over millions of years created a mixture of hydrocarbons which as a vapour formed natural gas while as a liquid it created oil. It is thought that the degree of heat and pressure together with the type of biomass determined whether gas or light oil was formed at higher temperatures and pressures, or more viscous heavier oil. The oil and gas migrated upwards from the source rock until it either escaped to the atmosphere or became trapped in the pore spaces of porous reservoir rocks such as sandstone, under a cap of clay or impermeable rock.

Table 3 shows the typical chemical composition of crude oil. The carbon and hydrogen atoms are arranged in complex hydrocarbon molecules that vary depending on the source of the crude. The heating value of crude oil when burnt (oxidized) is about twice that of coal.

Table 3. Typical composition of oil (% by weight) and heating value

Carbon %	Hydrogen %	Sulphur %	Nitrogen %	Oxygen %	Heating value MJ/kg
83–7	10–14	0.1–6	0.1–2	0.1–1.5	42–47

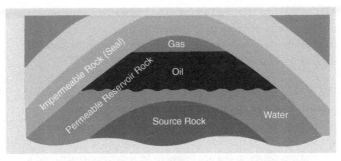

4. Schematic of an anticline trap forming an oil and gas reservoir.

A typical crude oil reservoir is formed by three rock layers: a source rock, a porous reservoir rock, and an impermeable cap rock. The source rock contains kerogen, which is a dense waxy organic material formed from decaying organic matter. Over millennia the complex hydrocarbons of the kerogen were cracked into lighter hydrocarbons under the heat and pressure of sedimentation. Typically at depths of up to 4 km and temperatures of 120 °C oil was formed, while at greater depths and higher temperatures of up to 200 °C gas was created. The oil and gas migrated upwards from the source rock and were trapped within the pores of the reservoir rock under the impervious cap rock. Figure 4 shows a simple schematic of an anticline trap creating a conventional crude oil and gas reservoir. The oil and gas are shown floating on a layer of water. The fraction of water contained within the reservoir rock is an important factor in deciding if it will be cost-effective to develop a well.

Exploration for oil and gas is undertaken in stages of increasing cost and accuracy. Initially a combination of aircraft or satellite remote surveys is used to identify suitable formations of sedimentary rocks. This is followed by seismic surveys that provide more detail of the geology to a greater depth. If these indicate that oil or gas is likely to be present in commercially recoverable quantities then this is confirmed by drilling exploration wells.

A conventional well is drilled vertically. A drilling derrick supports a drill string consisting of coupled lengths of pipe that are driven by a top mounted motor to rotate a drill bit. The cutting surfaces of the drill bit grind the rock into particles the size of grains. Drilling mud is injected to cool the bit and circulated to the surface where the rock residue is filtered out. The mud acts to keep the hole open and counteracts pressures of any gas or fluid encountered. If the well is to be used for production, a casing is inserted into the well and cemented into place to ensure that it remains open and the oil does not pollute any aquifers. The section of the casing that passes through the oil and gas bearing reservoir rock is then perforated using small explosive charges. Another internal tube is inserted to extract the mixture of oil and gas. A valve arrangement, known as a Christmas Tree, is installed at the top of the well to control the flow of oil and gas and direct it for treatment. Angled drilling techniques allow access to reservoirs over a wide area from a single surface location. Horizontal drilling starts vertically and then turns horizontally while multilateral drilling produces several branches from the main well.

A de-watering plant, located close to the well, separates the water from the oil using a cyclone. The water is often saline and not suitable for use in agriculture and so careful disposed is required, often by re-injection into the oil reservoir to increase the pressure. The gas is separated either for flaring or for processing and onward transmission. The resulting crude oil is then taken from the site either in tankers or by a pipeline.

Once a production well has been established, primary recovery is through the natural pressure within the reservoir forcing the oil to the surface. This may only produce 10 per cent of the oil in the reservoir. Secondary recovery is by re-injecting the water that is produced with the oil back into the reservoir through a second well and can result in a further 20 per cent of the oil in the reservoir being recovered. This also disposes of the water. Enhanced Oil Recovery techniques can result in 60 per cent of the oil in a reservoir being brought to the surface. Common techniques of Enhanced Oil Recovery are the use of steam injection to reduce the viscosity of the oil and increase its flow, injection of gases (e.g. CO_2) to increase the pressure in the reservoir, and injecting a mixture of water-soluble polymers that push the oil out of the pores of the rock.

Many of the easily accessible deposits of oil have been exploited and exploration has now moved to more demanding environments including in deep waters offshore. In shallow water, up to 150 m depth, jack-up rigs may be used. These are designed to be mobile and the drilling platform is supported on legs that are extended once the rig is in position. In deeper water, semi-submersible rigs or drill ships are either anchored to the seabed or use active thrusters to provide dynamic positioning. Offshore production platforms often include oil storage and are more likely to be larger structures fixed to the seabed or using some form of tension mooring. They can collect oil or gas from a number of wells.

Crude oil is a complex mix of hydrocarbons that is processed in an oil refinery into fuels including vehicle gasoline and diesel, heating oil, and jet fuel. Oil refining consists of three basic processes: separation, conversion, and treatment. To separate the oil, it is heated to around 400 °C and then passed through a distillation column where the light hydrocarbon gases rise to the top and are condensed to form the constituents of gasoline. The constituents of the medium weight fuels such as kerosene, diesel, and heating oil rise only part of the way up the column while the heavier molecules of fuel oil condense at the bottom of the column and are drawn off there. A distillation column may operate either at atmospheric pressure or under a vacuum. When the crude oil is separated by distillation, the fraction of each component of fuel that is produced is determined by the feedstock crude. Gasoline, diesel, and jet fuel are particularly valuable products although demand for them varies by season. It is often commercially attractive to maximize the production of particular fuels by converting the hydrocarbon molecules. This is done either by cracking long molecules into shorter ones by temperature and pressure, sometimes using catalysts or by combining shorter hydrocarbons. Finally treatment and blending processes are undertaken to remove undesirable impurities and ensure the quality of the fuel. Crude oil in a modern refinery may yield up to 50 per cent gasoline, 30 per cent diesel/heating oil, 10 per cent kerosene/jet fuel, and 10 per cent other products.

The oil industry can be divided into three areas: upstream exploration and production, midstream transportation, and downstream refining and distribution. It is usual for companies to operate predominantly in one area. Most upstream activity is undertaken by national oil companies, some of whom are members of the Organization of Oil Exporting Countries (OPEC). OPEC is an organization of fourteen oil producing nations that was formed in 1960 to manage the oil price and now controls around 44 per cent of worldwide crude production. The remaining 56 per cent of oil production is by national oil companies that are not members of OPEC, the well-known six major international oil companies of which the largest is ExxonMobil, and second tier more regional and smaller independent producers. Midstream transportation is undertaken by marine tanker and pipeline operators and, although some refineries are owned by companies also active upstream, many are independent.

The life cycle of an oil field typically follows Figure 5 although some very large fields in the Middle East were originally

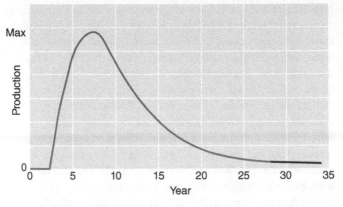

5. Typical life cycle of an oil field.

developed in the 1930s and are still producing. The recovery factor describes the fraction of oil that can be recovered. For the UK Continental Shelf oil fields this was around 43 per cent but it varies widely for individual fields. The recovery factor in gas fields is higher and can exceed 85 per cent.

In 1956 an American geologist, M. K. Hubbert, found that new discoveries of oil in the USA had peaked in the 1930s. Assuming a delay of around forty years between initial discovery of an oil well and peak output he predicted correctly that the production of conventional oil in the USA (excluding Alaska) would begin to decline from 1970. Similar effects have been noted in other countries with most discoveries of large fields being made before 1980. This has led to the concept of 'Peak Oil' and the prediction that the production of oil using low cost conventional drilling methods will begin to decrease before 2030. There appears to be good evidence to support this hypothesis including that the rate of decline of production from all oil fields at present producing is at least 4 per cent/year. This implies that an increase in oil production capacity equivalent to that of Saudi Arabia is required every three years simply to maintain current production. The recent improvements in unconventional oil extraction techniques have acted to increase production and some commentators contest the idea of Peak Oil. Also it has to be recognized that the data on reserves in some countries is unverifiable.

However, it is generally accepted that most of the large oil fields that can be accessed easily and at low cost using conventional extraction and processing methods have been identified and are being (or have been) exploited. Attention is now shifting to so-called transitional and unconventional oil. There are no strict definitions of these terms as they describe oils with a wide range of physical and chemical characteristics that require quite different production and processing techniques. One approach to their classification is by ease of extraction and processing, with conventional oils being the easiest to extract and process, and transitional and then unconventional oil being progressively more costly. The exploitation of transitional and unconventional oil resources has relied heavily on the development of directional drilling and hydraulic fracturing. The widespread use of these techniques in the USA is credited with significantly reducing its dependence on imported oil.

Transitional oil has a similar chemical composition to conventional oil but is extracted using new techniques. Tight oil is found trapped in geological formations with low permeability, such as shale, and is extracted by directional drilling and hydraulic fracturing. Ultra-deep oil is found offshore up to 10 km below the water surface and requires specialized drilling techniques.

Unconventional sources of oil include oil sands, oil shale, and shale oil or light tight oil. Oil sands consist of heavy complex hydrocarbon molecules closely bound to sand. There are large deposits of oil sands in Western Canada, Venezuela, and Siberia that yield bitumen or heavy oil. Extraction is either through surface mining or *in situ* techniques. When extracted with surface mining techniques, the oil sand is mined using large grab lines and transported to a local processing plant in very large trucks or by pipeline. Large quantities of hot water are used to reduce the viscosity of the oil and to separate it from the sand, sometimes using centrifuges. The resulting heavy oil or bitumen is then either upgraded into synthetic oil or blended with lighter crudes before it is transported to a refinery. The exploitation of oil sands is expensive and requires considerable heat energy that is often provided by natural gas. Thus the cost-effectiveness of producing crude from oil sands relies on a high price for the oil products and a source of cheap natural gas.

Oil shale is fine-grained sedimentary rock that contains between 5 and 40 per cent kerogen that did not experience the physical conditions necessary to transform it into oil or gas. Oil shale is surface mined and then processed by heating to around 450–500 °C in anaerobic conditions in a process of pyrolysis. This results in the long-chained kerogen molecules being split and when hydrogen is added lighter hydrocarbons are formed. An alternative *in situ* approach involves inserting heaters into the oil shale. There are very large deposits of oil shale distributed across the world but the process of extracting the oil is both expensive and energy intensive.

Shale oil, also known as light tight oil and not to be confused with oil shale, is found in shale beds and is extracted using techniques of horizontal drilling and hydraulic fracturing similar to those used for the production of shale gas. Shale oil and gas are often found together and the production of light tight oil relies on the pressure developed by associated gas to release the oil from the shale so that it migrates to the well. Compared to conventional oil production, a large number of wells are needed with the associated environmental impact and the techniques needed for extracting light tight oil have not been widely adopted in Europe. Production of light tight oil is increasing rapidly in the USA and accounted for 50 per cent of the US production of crude oil in 2017. Shale oil has only been extracted in large quantities for a decade and so there is limited experience of the life of the wells.

After its initial commercial exploitation in the late 19th century, the price of oil was remarkably stable at around US\$20/barrel (in 2016 dollars) for almost 100 years until political events in the Middle East led to short periods of prices of over US\$100/barrel in 1970 and again in 2012. The rapid increase in production of unconventional oil in the USA is generally credited with contributing to the dramatic reduction in the oil price in 2014, although the price of oil is heavily influenced by international political factors and the state of the world economy. Recently (2019) the spot price of oil has stabilized at around US\$60–80/barrel.

Liquid oil products are an extremely convenient form of energy and have a high energy density, are easy to transport, and can be stored easily. They are the main source of energy for transport either in gasoline (spark ignition) or diesel (compression ignition) reciprocating engines, or as kerosene in aviation gas turbines. However burning these hydrocarbons has major environmental impacts from their production of CO_2 as well as emitting particulates and oxides of sulphur and nitrogen. Large marine reciprocating engines often use fuels with higher levels of sulphur than are permitted for road vehicles or power stations and their pollution controls are limited. Hence they are a particularly important source of undesirable emissions but have yet to be regulated effectively. Electric cars are now becoming common and a transition pathway away from the use of fossil fuel for cars is emerging. However ships, heavy road vehicles, agricultural and forestry vehicles, and aeroplanes continue to be powered by oil and ways to operate them without producing CO_2, such as from electricity or hydrogen, are less clear. Also the use of oil as a raw material for petrochemicals is likely to continue.

Natural gas

The flexibility and convenience of gaseous fuels was recognized in the early 19th century when gas began to be manufactured by heating coal in an oxygen poor atmosphere in local gas works and was distributed through low-pressure pipe networks in towns and cities. Coal gas is a flammable mixture mainly of hydrogen and carbon monoxide, often with some methane, which was used initially for lighting homes and public spaces and later for cooking. Flammable gas can be manufactured through pyrolysis from many fossil hydrocarbon sources including coal and oil and is a by-product of many industrial processes such as steel making. Manufactured coal gas was replaced by natural gas starting in the 1940s in North America and the 1960s in Europe, but by this time electricity had superseded gas for lighting. The transition of gas supply from manufactured town gas to natural gas necessitated the replacement or modification of gas appliances although the pipe distribution network was retained.

Natural gas is the most versatile of the three main fossil fuels and in 2017 supplied around 23 per cent of worldwide traded energy and powered a similar fraction of the world's electricity generation. It consists of around 95 per cent methane (CH_4) with small quantities of higher hydrocarbon gases including ethane, propane, and butane as well as contaminants of nitrogen, carbon dioxide, and hydrogen sulphide. It is used widely in Europe and North America for domestic and industrial heating, electricity generation, and as a feedstock for chemical manufacture. Table 4 shows the typical composition and heating value of natural gas. Natural gas

Table 4. Typical composition of natural gas (% by weight) and heating value

Carbon %	Hydrogen %	Sulphur %	Nitrogen %	Oxygen %	Heating value MJ/kg
74	24	—	0.75	1	39

is usually measured in normal cubic metres, Nm^3, that is, cubic metres of gas at a temperature of 0 °C and pressure 101 kPa.

When burnt natural gas emits only 50 per cent of the CO_2 that would be emitted from coal for the same heat output and 73 per cent of that from oil. Burning natural gas results in very little SO_2 and less NO_x than many other fuels. It is particularly useful as a fuel for electricity generation using gas turbines either in large, high efficiency combined cycle units or in smaller open cycle turbines to provide flexible 'peaking' generation to compensate for the variable output of renewable generation. However, it has to be recognized that burning natural gas is a significant source of CO_2 and so its continuing use in large quantities for either heating or electricity generation is not compatible with a low carbon energy system.

It is thought that natural gas was created by the decay of buried organic matter over millennia in a similar manner to oil but at higher temperatures that led to the formation of simple gaseous hydrocarbons. It is often found when oil wells are developed and is then known as associated gas. Historically much of the associated natural gas found together with oil was flared as the costs of processing and transporting it to where it could be used could not be justified, but efforts are now being made to reduce this obviously wasteful practice. When natural gas is found on its own it is known as non-associated gas. Conventional gas reservoirs have defined extents and are usually accessed by drilling vertically through an impermeable seal rock and into a porous and permeable reservoir rock through which the oil and/or gas is drawn to the well and initially lifted to the surface by the natural pressure of the reservoir.

However, increasingly gas is being extracted from poorly defined continuous reservoirs in layers of shale, pockets of sandstone, or coal seams. The shale acts as both the source and reservoir rock. Shale has extremely small pore size and is impermeable to gas and oil flow unless fractured either naturally or artificially. Unconventional gas extraction involves horizontal or directional drilling followed by hydraulic fracturing to release the gas from rock whose natural permeability is too low to allow gas to flow in useful quantities.

The geology of natural gas resources is illustrated simply in Figure 6. This shows conventional non-associated and associated gas that would be accessed by vertical drilling as well as unconventional sources that require horizontal drilling and hydraulic fracturing. Typically coal bed methane is found at depths up to 1 kilometre while shale and tight gas is found at between 2 and 5 kilometres. In 2017 about 60 per cent of natural gas production in the USA was extracted using unconventional methods.

For the production of shale gas, a well is initially drilled vertically and then turned to follow the layer of shale. Well casing is inserted, the well sealed with cement to avoid contamination of aquifers, and small explosive charges used to puncture the casing within the shale layer. The rock is then fractured by injecting large volumes of water at high pressure to increase its porosity, sand is

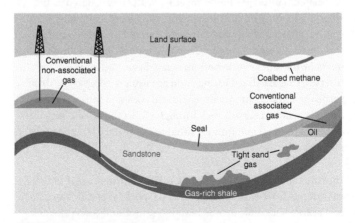

6. Schematic geology of natural gas.

inserted to keep the resulting fissures open when the water is pumped out, and chemicals applied to assist the flow of gas. The gas then flows through the fractured rock.

The development of natural gas (and oil) wells through hydraulic fracturing is widespread in the USA but is not permitted in many countries. Hydraulic fracturing (fracking) has been used to augment the flow from conventional, vertical wells since 1949 but in Europe its proposed use on long horizontal wells to extract shale gas has generated considerable public opposition. The objections to fracking are based on concerns over the contamination of aquifers from the chemicals used, creation of small earthquakes, and the general environmental impact of drilling in rural areas as well as the increased road traffic that is required. Some environmentalists see fracking, with the transition from highly carbon-intensive coal to less carbon-intensive natural gas as a fuel for generating electricity, as a distraction from measures to decarbonize the energy system.

Gas from a well is processed to improve its quality either at the site or at a gas terminal. The processing stages include separating any oil and hydrocarbon condensates, removing water, and drying and cleaning the gas of contaminants such as hydrogen and carbon dioxide. The dry pipeline quality natural gas is then suitable for use as a chemical feedstock or injection into the high-pressure transmission system.

Gas can be used to generate electricity by burning it in a steam-raising boiler but more commonly it is used in a Combined Cycle Gas Turbine (CCGT). Figure 7 shows a schematic of a CCGT. Gas is used to power a gas turbine that turns a generator. The waste hot gases from the gas turbine are passed through a heat exchanger to raise steam for a second turbine and generator. When the steam has passed through that turbine it is cooled and condensed. In this way, two thermodynamic cycles are operated in tandem to increase the efficiency of electricity generation to more than 60 per cent.

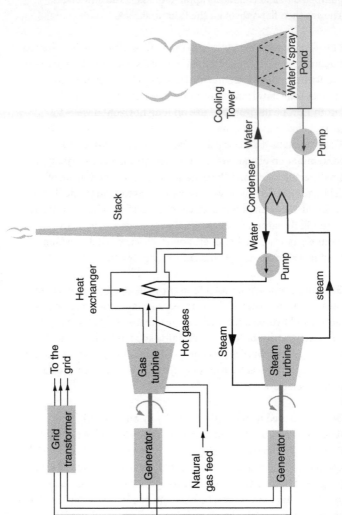

7. Combined cycle gas turbine.

Generation of electricity from natural gas is less carbon intensive than from coal or oil but still results in emissions of around 350 tonnes CO_2/GWh when used in a CCGT unit. Fugitive emissions of natural gas are also of concern as methane has a global warming potential approximately thirty times that of CO_2. Natural gas is seen by some commentators as an important transition fuel to a low carbon energy system. However others express concern that the exploitation of any hydrocarbon source diverts attention away from the need to create a truly low carbon energy economy.

Figure 8 shows the natural gas system of GB and its various sources. Gas supplies come from large, but rapidly depleting, fields on the continental shelf as well as pipeline interconnectors from neighbouring countries. There are several facilities to receive and regasify Liquid Natural Gas (LNG). LNG is refrigerated and compressed gas shipped internationally in special ships. The onshore gas fields in GB are small and hydraulic fracturing has only been undertaken experimentally. The gas from all these sources is brought to a gas terminal and then processed and blended before being injected into the National Transmission System (NTS). This network of high pressure pipelines is used to supply gas to large power stations and industrial users. Pressure is maintained in the NTS by compressors located along the pipelines. The compressors can be driven by gas turbines but are now usually electrically powered. Gas is distributed to users through a succession of pressure reducing stages until it is supplied to domestic customers at less than 75 mbar (approx. 30 inches water gauge).

The pressure of the NTS is allowed to vary throughout the day between 45 and 70 bar and this variable quantity of gas (known as linepack) provides important energy storage for day-to-day operation. There is some dedicated gas storage either underground or in pressurized containers on the surface although GB is unusual in having only a few days of gas storage. When gas from the UK continental shelf field was plentiful, dedicated gas

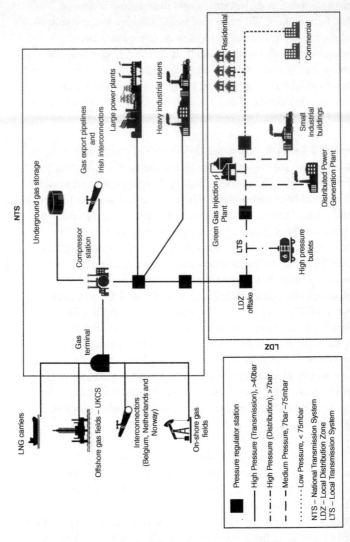

8. Simplified schematic of the GB natural gas system.

storage was not needed and now in a liberalized market it is hard to recover the high costs of investment in storage that is used only rarely. Opinion is divided as to whether GB has an appropriate level of gas storage to maintain supplies during coincident periods of cold temperature and supply disruptions. Large industrial users and electricity generating stations usually have lower cost contracts with gas suppliers that allow them to be disconnected at times of gas shortage. These interruptible contracts reduce the need for storage.

The national gas system is an important asset and in GB on a winter's day it transmits more than four times the energy that passes through the electricity network. Thus there are considerable efforts under way to adapt it for use in a future low carbon energy system. Biogas from the anaerobic digestion of biological material or organic waste is cleaned, CO_2 removed, and the methane injected into the gas network. Methane is also produced from landfill although this is usually used to fuel small electricity generators on the site. It is argued that burning methane produced from biomass is essentially carbon neutral.

Alternatively, as a gas is much easier to store than electricity, surplus electrical energy from wind or photovoltaic generation can be converted into hydrogen by electrolysis of water. When hydrogen is burnt it produces only water. Different national natural gas systems have limits to the amount of hydrogen that they can allow, that range between 0.1 per cent (GB) and 12 per cent (Holland) by volume for various, often historical, reasons and work is under way to establish the real technical limitations. More radically, trials are in progress to convert sections of the gas network to hydrogen. Many of the original gas low pressure pipes have already been fitted with polyethylene liners that are suitable for transporting hydrogen. However, changing the gas from methane to hydrogen would require all gas appliances to be adjusted or modified.

Chapter 3
Electricity systems

Electricity is not a primary fuel but a means by which energy is distributed and so is an energy carrier or vector. Once it has been generated and transported to its point of use, electricity is versatile, controlled easily and has very low environmental impact. It can deliver energy for lighting and heating, and supply motive power with high efficiency. It is essential to power electronic equipment and so is the preferred medium by which energy is transported in modern society.

However electricity suffers from the key disadvantage that it cannot be stored in large quantities cost-effectively and so it is necessary to match the supply of electrical energy to the demand at all times. The two essential functions of an electricity system are to take power from the generators and distribute it to consumers, and to balance the supply and demand of electrical energy. A secondary objective is to control the voltage but this is easier and cheaper to accomplish.

In a modern electricity system, multiple generators are used to meet the load. Some types of generator have high capital but low running cost and so are used much of the time, so-called base-load generation. Other generators with low capital but high running costs only operate at times of peak load. Extensive networks of either overhead lines or underground cables interconnect the generators and transport the electricity to the loads.

The first small electricity generating stations and distribution networks serving the public were installed in New York and London in the 1880s. The generators were driven by coal-powered steam engines, although waterpower was also used to power small private electricity generators around that time. These early power stations provided lighting for public spaces and residences. For some years electricity was in competition with manufactured town gas, until gas was eclipsed by electricity for lighting. Public electricity systems then spread rapidly, initially supplied from large numbers of small generating stations each providing power to a limited area before large central generators were installed to power much wider regions through interconnected transmission grids.

Electricity systems are characterized by their voltage and current. A useful analogy is with a fluid system; voltage is analogous to pressure and current is analogous to the flow rate of the fluid. Some early electricity legislation was based on gas law and used the term 'pressure' to describe voltage. The power that is transmitted through a fluid system is the product of pressure and flow rate, while in an electrical system power is the product of voltage and current. Just as it is cheaper and creates lower losses to transmit fluid power at high pressure and low flow rates, it is usually more cost-effective to use a high voltage and low current to transmit electricity. This is true particularly if air can be used to insulate the conductors in overhead lines rather than the solid insulation required in underground cables. Hence bulk electricity is transmitted at high voltage (e.g. 400 kV) and large but not excessive currents (e.g. 2 kA).

In the early days of electricity supply it was unclear whether electricity should be generated and transmitted at ac or dc (see Box 3 and Figure 9). The advantages of ac are that the voltage

Box 3 Direct and alternating current

Electricity can be generated either at alternating or direct current (ac or dc). Figure 9a shows direct current, which is a constant current. Figure 9b shows alternating current, which is a sine wave oscillating between a positive and negative peak value every 20 milliseconds (for a system operating at 50 Hz). It can be seen that with alternating current the wires carry the maximum current only occasionally so making poor use of the conductors. Therefore three wires are used with three sine wave currents delayed from each other by 6.66 milliseconds. This is 3-phase alternating current and it is almost universally used for electricity transmission and distribution. For reasons of simplicity and safety in some countries (e.g. the UK) electricity is supplied to single houses with only one sine wave, but for larger loads 3-phases are always used.

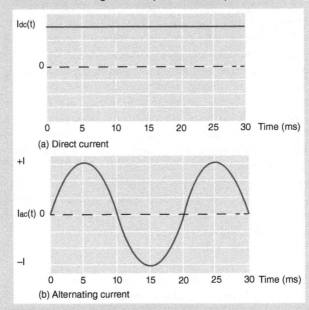

9. Direct and alternating current.

can be changed easily through transformers and so power can be transmitted at high voltage and relatively low current over long distances with low losses before the voltage is reduced at the point of use. Three phases are used as this makes best use of conductors and transmission corridors. Another important reason for the use of 3-phase ac is that an important type of electric motor, the induction machine that is the workhorse of industry, needs ac to operate. Hence after some years of both technical and commercial competition between ac and dc the world settled on 3-phase ac, 50 Hz in Europe or 60 Hz in much of the Americas.

The choice of frequency of 50 or 60 Hz was made in the early 20th century and was a finely balanced engineering judgement based on the ability to transmit power long distances and to operate incandescent lights without visible flicker. Japan has a power system that operates partly at 50 Hz and partly at 60 Hz. Europe operates at 50 Hz and North America at 60 Hz. One industrial part of northern England was operated at 40 Hz for many years, unable to be connected directly to the rest of the country, as the engineer responsible for the design considered 40 Hz to be preferable to 50 Hz. For some specialist purposes different frequencies are used, for example 400 Hz is often used in aircraft and $16\frac{2}{3}$ Hz for railway traction. Studies have also been undertaken to determine if 6-phases might be preferable to 3-phases for long distance transmission.

As late as 1980 there were still legacy dc systems in some cities and I once stayed in a hotel room that had both ac and dc connections to power two independent lights and fans for use when either system was unavailable during rota load shedding. Connecting the two systems could have caused an electrical fire.

Information and Communications Technology (ICT) systems require low voltage dc for their operation and with modern power semiconductors the voltage of direct current can be changed. Hence studies and demonstration projects continue to investigate whether dc should be used to distribute power in particular circumstances such as in large commercial buildings or computer server farms. However these small trials have not threatened the dominance of ac.

In the early days of public electricity supply (up to the 1930s) it was common for electricity to be supplied by a local independent power station serving a restricted geographic area such as a town or part of a city. The benefits of interconnection through a grid were recognized but this posed a number of technical problems, including the stability of the system. Learned papers were written in the late 1920s describing the potential danger of interconnecting generators together into a grid but these concerns were not well founded and national or state wide grids were established. Now a single grid connects much of mainland Europe and large synchronous grids connect multiple states of the USA.

Established electricity systems

Figure 10 shows a typical large thermal ac power supply system. The generators are in large central power stations that are usually located some distance from the loads, where the electrical power is consumed, in order to minimize the environmental impact of emissions from burning fossil fuel and to have good access to fuel and water for the turbines. The transmission system transports bulk supplies of electricity while lower voltage distribution networks take the electricity to the loads.

The fuel for steam turbines has often been low cost coal supplied by rail although oil or gas can also be burnt in the boilers. Increasingly gas turbines powered by natural gas are used and these need good access to the high pressure gas network. The location of hydro turbines is determined by the water resource.

Large volumes of water are needed for cooling steam turbines and the available water supply can be an important constraint on their

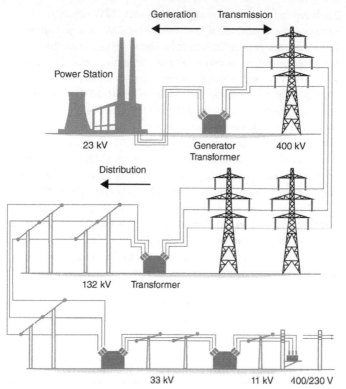

10. Interconnected electricity system.

location and operation. The environmental impacts of using water to cool thermal generating plants is a topic of increasing concern for power system operators and environmentalists. During extreme heat waves, some steam driven generators (nuclear or fossil fuelled) that take and return cooling water to estuaries or large rivers have to shut down when temperatures of the watercourses exceed agreed environmental limits. In some countries the direct use of water extracted from rivers and even the sea and returned at a higher temperature is not permitted and cooling towers are required. These are expensive and reduce the efficiency of the generators.

Each generator, rated typically at up to ~500 MW capacity, operates at a comparatively low voltage (~25 kV) and so with very large currents. It is difficult to use a higher voltage than this in a rotating generator but very thick conductors are needed to carry such large currents. Therefore a transformer is sited close to each generator to increase the voltage to the 400 kV or 765 kV of the interconnected transmission grid but reduce the current. This high voltage network transports the power, often over considerable distances, to the load centres. The voltage is progressively reduced in the distribution network through a series of transformers and the circuits then become radial. The voltage is held approximately constant at each level of transformation. Details of the voltage levels differ country by country but the distribution networks of European countries with high population densities all follow similar principles. In parts of the USA with lower population densities, higher distribution voltages are used with more frequent, smaller transformers.

Table 5 shows the voltage levels used in different countries although these are only approximate as practice varies, often depending on historical precedent.

Table 5. Typical voltage levels of electricity networks

Network	UK practice	European practice	US practice
Supplies to households	230 V single phase	230 V single phase or 400 V 3-phase	120 V single phase or 208 V 3-phase
Low voltage distribution	400 V	400 V	120–600 V
Medium voltage distribution	11–132 kV	10–150 kV	2.4–34.5 kV
High and extra high voltage transmission	275–400 kV	220–400 kV	46–765 kV

The conventional structure of a modern electricity supply system, shown in Figure 10, has a small number of large central power stations. These supply a high voltage transmission network of overhead circuits on lattice steel towers (pylons). The lower voltage distribution circuits may be overhead lines on wood or concrete poles in the countryside, and underground cables in urban areas (Box 4).

Box 4 Historical development of the British power system

Electric power distribution in many towns and cities began with systems radiating from a small power station. Power was generated and distributed either at ac or dc at the low voltage at which it was used (i.e. 100–200 V).

Larger more efficient generators were needed but the low voltage of the distribution system limited the area that could be supplied and hence the size of the generators. Therefore, new medium voltage networks were interposed between new larger power stations and the low voltage systems that supplied consumers. Each section of low voltage network, serving a small area, was supplied from the medium voltage network through a transformer.

The voltages used were increased several times through the 20th century. In the UK, initially 33 kV was used with power stations of about 30 MW. Then, in the 1930s, a mainly overhead 132 kV grid began to be constructed to interconnect these stations. After 1945 power stations of about 200 MW were connected to the 132 kV grid. In the 1950s and 1960s, the distribution networks were standardized on 3-phase ac at a few preferred voltages. All low-voltage supplies were made 240 V ac or 415 V 3-phase (later reduced to 230 V and 400 V). Medium

(continued)

The load varies throughout the day and season by season.
Figure 11 shows typical load curves of the GB system. Maximum
load is experienced in the early evening during the winter and
minimum load in the early morning in the summer. There is a
ratio of more than two to one between minimum and maximum
load. Such variations of load are not uncommon in a national
power system but in some countries where air conditioners are
used the load peaks in the summer.

The instantaneous balance between generation and load is
achieved by adjusting the output of the generators to meet the
demand. The size of the system allows large, high efficiency
generators to be used. The steam turbines powered by fossil fuels
or nuclear fission have restrictions on how rapidly they can change
their output (known as their ramp rate) while hydro turbines are
more flexible and can alter their output quickly. Small open cycle
gas turbines are flexible and can rapidly follow changes in load but

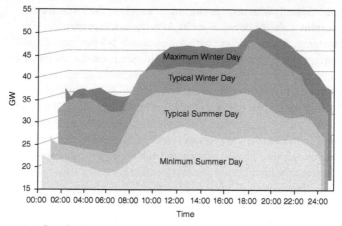

11. Load on the GB power system.

large combined cycle gas and steam turbine units have limits to their ramp rates.

Faults, such as lightning strikes, are inevitable on any electrical power system and have to be considered in its design and operation. The possibility of a generating unit tripping limits the maximum size of generator that can be installed and also requires several units to be operated continuously at below their maximum output so that they can quickly increase their output and compensate for any loss of generation from the tripped unit. Operating a thermal unit at less than full output reduces its efficiency and increases cost. Hence the requirement for maintaining operating reserve is an important commercial consideration.

The high voltage transmission grid is very reliable with two or three redundant pathways from the generators to the distribution systems in service at any time. The grid is equipped with protection systems that detect any fault and automatically isolate the faulty component to allow the system to continue to operate satisfactorily with no interruption of supply to the loads. As the current is shared between multiple pathways, transmission systems are very efficient with losses of only 1–2 per cent of the energy that is transported. The distribution networks use single or duplicate radial circuits and so are less reliable and when faults occur they have to be repaired. The reduced redundancy of distribution networks leads to higher losses, typically of 6 per cent of the energy transported. These distribution losses are inevitable and cannot be reduced without considerable investment in larger overhead lines, underground cables, and transformers. Regrettably, distribution systems in some countries also suffer from 'non-technical' losses such as theft of electricity or non-payment of bills.

Although it must operate as an integrated whole, it is common for a power system to be divided into generation, transmission, and distribution for administrative purposes and sometimes for ownership. In a traditional large state owned integrated utility all aspects of electricity supply come under a single organization but the different skills required for each aspect mean responsibility is given to different departments. The generation of electricity requires mechanical engineering expertise, while the high voltage transmission network is very sophisticated requiring complex control but with a limited number of very expensive components. In contrast the distribution system contains a very large number of simpler elements and its operation involves contact with a large number of consumers.

Between 1945 and 1990 in Europe it was common for all aspects of electricity supply to be the responsibility of large, vertically integrated, and often state owned national utilities. Since then there have been progressive moves to split up and privatize these utilities and introduce competition in generation and energy retailing. Privatization was driven partly by the belief that large state owned monopolies were inherently inefficient and partly to remove from governments the obligation to provide the very large capital investments needed to develop electricity supply systems. The generators were privatized to establish competition between them and the role of energy traders, sometimes called retailers, was introduced. Energy retailers or suppliers do not own physical assets but buy electrical energy in bulk from generators and sell it to consumers through real-time markets and other commercial instruments. The transmission and distribution networks are natural monopolies and continue to be operated by licensees under a national regulator. In the USA, a mixture of publicly owned and investor owned utilities supply electricity, although sometimes distribution is the responsibility of municipal authorities while in some rural areas power is distributed through rural electricity cooperatives.

Contemporary developments in generation

Although the privatization of electricity supply alters the ownership and commercial operation of a power system, the underlying physical characteristics and technical operation remain unchanged. Thus the main functions of generation, transmission, and distribution and the overall technical architecture of the electrical power systems have remained essentially unchanged since modern electricity systems were established in the 1950s. However each of these functions is now facing radical technical change. This is due to a combination of the need to reduce CO_2 emissions and other environmental impacts and our dramatically increasing ability to gather and process large amounts of data in real time and so use advanced control techniques. A further potential disruption to traditional electricity systems is the anticipated rapid cost reduction of storage of electrical energy in batteries.

The Paris Climate Change Agreement is the latest international effort to address the effect of greenhouse gases on the environment and is intended to limit the increase in global average temperature to well below 2 °C above pre-industrial levels. It was negotiated within the United Nations Framework Convention on Climate Change (UNFCCC) in 2015 and is due to come into effect in 2020. Each nation proposes, implements, and reports its own programme to meet this target. Electricity supply typically creates between 25 and 40 per cent of a nation's emissions of CO_2 and many countries recognize that decarbonizing their electricity system is an important and achievable way of helping reach their environmental targets. A realistic initial target for the emissions from a decarbonized power system is between 50 and 100 tonnes of CO_2 for each GWh of electricity supplied.

Thermal generators, particularly if fuelled by coal, are major sources of carbon dioxide and replacing their output by electricity generated from low carbon sources (such as nuclear energy and renewables) is an obvious way to reduce emissions of greenhouse gases. Table 6 shows the carbon intensity of various forms of electricity generation. Burning coal to generate electricity creates a large amount of CO_2 and is not compatible with ambitions to limit emissions. Raising the temperature and pressure of the steam can increase the efficiency of a turbine generator but the

Table 6. Carbon intensity of electricity generation

Fuel	Tonnes of CO_2/ GWh of electrical output
Coal (pulverized fuel boiler)	920
Oil (not used in GB)	633
Natural gas (combined cycle)	349
Great Britain generation portfolio in 2018 (including nuclear and renewables)	208
Target for a decarbonized electricity system	50–100

Data from Digest of UK Energy Statistics. https://www.gov.uk/government/collections/digest-of-ukenergy-statistics-dukes Accessed 1/5/19 © Crown copyright—content available under the Open Government Licence v3.0.

materials used in the boiler limit these increases and hence the improvement in efficiency and reduction in CO_2 emissions that can be achieved. Capturing the CO_2 and either sequestering it underground or in the sea, or using it for the manufacture of chemical materials, remain laudable aspirations but progress in the development of these technologies is slow.

The fundamental laws of physics and the maximum temperature that the materials can withstand limit the efficiency of any single cycle thermal electricity generator (steam turbine, gas turbine, or reciprocating engine) to a maximum of between 30 and 40 per cent with 60–70 per cent of the heat being rejected to the atmosphere. There are two common approaches to make better use of the energy in the fuel when generating electricity: Combined Heat and Power (CHP), sometimes called co-generation, and combined cycle operation.

CHP increases the efficiency with which fuel is burnt by taking the rejected heat and using it either for district heating or in industrial processes. The overall efficiency of a CHP scheme can be around 75 per cent with some 30 per cent of the output energy being electricity and 45 per cent useful heat. CHP generators providing electricity and steam or hot water to district heating networks are common in northern European cities that have a high demand for heat throughout a long winter. Combined heat and power stations that supply district heating networks are now gaining acceptance in more temperate climates for both district heating and cooling. The rejected heat energy can be used for cooling by absorption chillers.

Combined cycle operation improves the efficiency with which electricity is generated by combining a gas turbine generator with a waste heat boiler that creates steam (see Figure 8). The steam is then used to power a second generator. Large modern combined cycle plants have an efficiency of electricity generation of 60 per cent with low emissions of SO_2 and NO_x.

The need to reduce emissions of CO_2 and the support of many governments for low carbon generation is leading to rapid change in the operation and control of national electricity systems. Existing nuclear plants as well as the commercially available new designs are large central units that are operated at constant output to maintain stable conditions in the reactors and steam turbines. Even if their technical operation allows their power output to be varied, the high capital costs of nuclear generation can only be recovered through generating electricity and so there is every incentive to operate these plants throughout the year at their full output. Thus nuclear generators usually operate at a fixed output and so are described as 'inflexible'.

Both wind power and solar PV generation have benefited from government support in many countries for some years and their capacity has grown rapidly, often through guaranteed premium prices for the electricity generated. Both technologies have recently seen considerable cost reductions that are now encouraging proposals from project developers that rely much less on public funding. The power from renewable sources depends on environmental conditions and is not under the control of the operator; the power generated varies with the weather and can only be predicted with some uncertainty. Thus their output is described as 'variable'.

Electrical energy cannot be stored cost-effectively in large quantities and in an ac power system the frequency must be maintained within strict limits. If the electricity being generated at any time exceeds the load connected then the system frequency rises, while if there is more load than generation the frequency drops. This balance of energy supply and demand must be maintained second by second and is achieved by connecting and disconnecting generators as they are needed and automatically adjusting the output of some of the operating generators to follow the ever changing load. For a generator to be able to adjust its output to follow the load it needs a source of stored energy, in either fossil fuel or water stored at a height. The combination of inflexible nuclear generation and variable renewables is a challenge for controlling the frequency of a low carbon electricity system.

Thus, low carbon generation has technical and commercial characteristics that are new in generating systems, and as more and more renewable generation is used these are starting to have a significant impact. Neither nuclear nor renewable generation contributes directly to matching supply and demand. Moreover, both reduce the load seen by the flexible fossil generators that are able to change their output, so reducing their opportunities to operate and gain revenue.

Nuclear and renewable generators have high capital cost but very low operating (fuel) costs. In a centrally operated system they appear low in the merit order, in which generators are called to operate starting with the cheapest (see Box 5 and Figure 12 for an explanation of the merit order). In a market based system they will bid as low as is necessary in order to operate for as many hours in the year as they can. In some countries renewable generation receives a guaranteed price for the electricity it generates through a Feed-in-Tariff and has preferential access to

Box 5 The generation merit order

The merit order, shown in Figure 12, is the classical way of representing the operation of a generating system. It shows the generators of a power system ordered by their operating (short run marginal) cost that is made up of fuel and variable maintenance and staff costs. For minimum cost of operation, the generators are called on to supply the load in order of ascending marginal cost and the price of electricity is the cost of operating the most expensive unit needed. As demand varies throughout the day and year, units are called on to operate following this principle.

(continued)

Box 5 Continued

Low carbon units, renewables and nuclear, have very low marginal costs and so always operate if the generators are available and the weather permits. When low carbon plant operates it is positioned at the extreme left and the rest of the merit order is translated towards the right. This has the effect of reducing both the average price of electricity and the number of hours of operation that fossil plant can run. Plant with energy storage (fossil or hydro) is still required to follow the net load (the consumer load minus low carbon generation) and as reserve for plant breakdowns. The effect of adding low carbon generation to the power system is to reduce the opportunities for conventional plant to operate but the need for flexible reserve plant increases. These new operating requirements are incompatible with the traditional practice of rewarding generation by the electrical energy generated (MWh) and increasingly generators have to be rewarded through capacity payments. Capacity payments are made to ensure generators are available when needed even if they are not called on to operate.

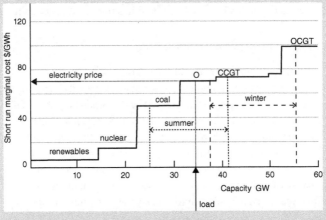

12. **Merit order, showing operating point (O).**

the network. Thus in many power systems there is an increasing fraction of low carbon generation that will always operate, when weather conditions permit, at low marginal cost. This reduces the cost of electricity to the consumer and makes the supply of electrical energy (MWh) less valuable but the ability to produce power (MW) when needed much more useful, for example at times of high load but low wind and sun.

Traditionally, generators have been rewarded by the amount of electrical energy (MWh) they produce, and as nuclear and renewables produce more energy, the fossil generators can operate for a shorter time. Hence fossil generators have less opportunity to recover their capital costs. It is becoming increasingly apparent that in a low carbon power system, where nuclear and renewable generation are encouraged, the role of the remaining fossil fuelled plant is less to supply energy (MWh) and more to balance supply and demand power and provide flexible power (MW). If these generators are to be rewarded through the sale of energy (MWh) in a market then very volatile electricity prices can be expected. There have already been instances in some electricity markets of both very high (at times of shortage) and negative prices (at times of surplus) of electrical energy. The uncertainty of its future role makes financing new base load, fossil generation plant difficult and several traditional electricity generating utilities have restructured recently in recognition of the changing commercial environment.

The requirement for flexibility is illustrated by the well-known 'duck curve' of California (Figure 13). This shows the net load, that is, the load seen by the large utility generators, over a day. The load in 2013, before the widespread installation of PV systems, was essentially constant throughout the day before rising to an evening peak caused by lighting and air-conditioning. However the widespread installation of PV by 2020 is expected to lead to a reduction in the load seen by the large utility generators during the day (when consumers are supplied

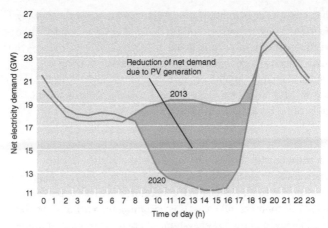

13. Impact of photovoltaic generation on the net electricity demand in California.

by solar PV farms and their roof-top PV systems) and then a
pronounced increase in the evening as the sun goes down. It may
be seen that the utility generators provide less energy throughout
the day but then must supply a steep ramp in load power. There
is a similar effect in GB at midday during summer weekends
when the consumer load is at a minimum and there is a high level
of solar PV generation. The net demand is reduced in the middle
of the day but it is necessary to keep thermal generators spinning
so that they can rapidly pick up load when the sun drops in the
evening or if clouds appear.

It has been conventional practice for an electric power system
to be designed and operated always to supply the load, unless
unforeseen breakdowns occur of a generator or in the
transmission network. Some failure to supply the load for a few
hours per year must be anticipated and accepted as faults are
inevitable and a completely reliable system would need very high
reserve capacity and so be very expensive. Traditionally, some
generators are kept in reserve, and others are operated at below

their maximum output so that they can quickly compensate for lost generation when faults occur.

Providing reliability is a major cost of any power system and there are considerable savings to be made by not always requiring the generators to follow the load but reducing the load at times of limited generation or restricted transmission capacity. This approach of reducing load when major faults occur or an unexpected generation deficit is anticipated is common in established power systems. Load is reduced either by increasing the price of electricity or issuing direct instruction to large consumers who have previously agreed to reduce their demand at times of a deficit of generation in return for cheaper power. A more extreme example of load control is the rota load shedding in countries with a structural deficit of either generating or grid capacity. The flexibility available from this Demand Side Management is becoming increasingly important in decarbonized power systems when many of the generators will be powered by renewable energy and so their output cannot be controlled or predicted with certainty.

The generators of a power system are very large masses rotating very rapidly at exactly the same speed and they store considerable spinning energy. This spinning energy controls the speed at which the system frequency changes when a fault occurs and there is a deficit or surplus of generation. This is sometimes known as the 'inertia effect'. Much of the new renewable generation is connected to the power system through static power electronic converters, which do not have the large spinning mass of traditional plant. As renewable generation is added it becomes necessary to disconnect the conventional spinning generators to maintain the supply–demand balance. With an increasing fraction of renewable generation connected through inverters, the system frequency changes more rapidly. This effect is sometimes described as the system becoming lighter and is of increasing concern to system operators.

Contemporary developments in transmission

The changes of the types of load and generation in a decarbonized power system are creating new requirements for the bulk transport of electricity. The need to deliver electrical energy through the transmission system will increase as road vehicles and other low carbon loads are powered by electricity but will be reduced by larger amounts of distributed generation being connected close to loads in distribution networks. These changes may counteract each other and the energy transported over a year through a transmission network may remain approximately constant. However, increased capacity is likely to be needed to accommodate occasional peak flows of power. This reduction in the intensity of use (or capacity factor) of the transmission system is similar to that being experienced by fossil fuelled generation.

Traditionally each country or state developed its own transmission network and had limited interconnection with its neighbours. Interconnectors with other countries were intended mainly to provide mutual support in the event of faults occurring within a national network. Recently, with the increasing capacity of generation from variable renewable sources, it has become advantageous to trade electrical energy across national boundaries and to take advantage of differing environmental conditions, that is, the strength of the wind or sun.

Also new ICTs are emerging that allow much greater visibility and knowledge of the state of the power system in real time as well as improved control. This is known as the emergence of the 'Smart Grid'. The transmission network is the backbone of the present electricity system and, although covering a very large area and able to transport very large amounts of power, it consists of a relatively small number of substations and circuits. Thus it is economic to invest in sophisticated technologies and practices in the transmission systems that are too expensive to be used in the much greater number of circuits of the distribution system.

The established method of transmitting large quantities of electrical energy is through 3-phase ac but there is a limit to the distance that ac power can be transmitted. For ac overhead lines this limit can be overcome by adding additional equipment along the route but for long undersea cables dc is required. The development of large wind farms, either offshore or in remote areas of a large country such as the USA or China, creates a need to transport electricity from sparsely populated areas to the load centres of cities. This has led to the need for long distance transmission and in particular the use of HVdc.

A dc system works by rectifying the 3-phase ac to dc at the sending end converter station, transmitting the dc over two conductors and inverting it back to 3-phase ac at the receiving end. High power semiconductor devices make the conversion from ac to dc and back, and control the flow of power. The terminal converter stations of a dc scheme are more expensive than ac substations but the overhead line or cable is cheaper, being only two conductors rather than three. The choice of whether to use ac or dc at high voltage is usually made on cost with dc transmission being considered for overhead line circuits of more than 600 km but for submarine cable circuits as short as 50 km for 400 kV and 150 km for 220 kV circuits. With the increasing use of high voltage, high current semiconductor devices, converter stations and their controls are becoming cheaper and more reliable, so making dc more attractive for shorter distances. Direct current transmission is also used to connect power systems that cannot be operated in synchronism for technical reasons. For example the 50 Hz and 60 Hz systems of parts of Japan are connected using dc links.

Transmission systems are extremely important for the secure supply of electricity and consist of a relatively small number of circuits and substations. Thus even when telecommunications were less developed, the status of the main points of the network was constantly monitored and telemetered back to manned control rooms. The telemetered data typically included current and power flows in the circuits and voltages at various points, and the system frequency was sent back to the control room every five seconds or so. Over the last twenty years it has become possible to use the very precise timing information available from the worldwide global positioning system to measure the phase angle of the ac voltage or current wave at various points in the system. This is particularly useful to examine the behaviour of the network during faults and to anticipate instability. These Phasor Measurement Units give much more detailed information of the state of the power system and when combined with detailed offline simulation studies allow the power system to be operated with greatly increased confidence.

The security and integrity of the communication system used to monitor and control transmission networks is of great concern to power system operators and cyber security receives constant attention. Usually the control and monitoring systems communicate through separate communication systems that are distinct from the other less critical information networks of the power companies and are heavily protected against intrusion.

Contemporary developments in distribution

Traditionally, the electricity distribution system has been the simplest, if most extensive, part of the power system. Its function was to take electricity from the transmission network and deliver it to consumers through simple passive, one-way radial circuits. The behaviour of loads was well known and could be predicted in aggregate with a high degree of certainty. The original distribution networks were carefully designed to connect the load at minimum cost. Great use was made of the diversity of the load, that is, the principle that although an individual house might have a peak demand of up to 10 kW, 100 such houses with the occupants behaving in their usual manner would have an average peak demand each of only 2 kW as people used appliances at different times. Thus the low voltage connection to an individual house was designed for 10 kW but the medium voltage supply to the 100 houses was designed for only 200 kW. This predictable behaviour of consumers can no longer be assumed as new types of load and generation are connected and loads can be controlled easily either automatically or from a Smart Phone.

When the distribution network was built, well before the era of mobile phones, the cost of communications was high. There was very little monitoring or automatic control installed to disconnect faulted sections of the network other than fuses or simple circuit breakers. There is still no continuous monitoring of many low voltage circuits and when a fault occurs disconnecting a single house or small section of the network the distribution network operator is unaware until the affected customers phone a call centre to report having no lights. As the low voltage network is radial the circuit must be repaired before supply can be restored. If a larger section of the distribution network is lost then switches, often remotely operated from a control centre, are used to provide an alternative path but with a delay.

For many years the electricity consumed by a domestic customer was measured using a simple electro-mechanical meter that was read manually only occasionally and a paper bill issued. There was no visibility of how the loads varied over time and network design was based on a small number of profiles of the load of typical consumers. Such simple procedures are still followed, but many countries have initiated programmes to install smart meters. These measure and record energy use typically every half-hour although they report their measurements to a central database typically daily. Smart meters allow much greater visibility of electricity use and can be used to devise time-of-use tariffs that encourage customers to use electricity when it is cheaper.

However either direct control of non-essential loads or a varying price of electricity can disrupt the natural diversity on which the distribution network was designed.

From around the year 2000, distributed generators began to be connected to the distribution network in increasing numbers. These generators, which were not anticipated when the circuits were originally designed, are typically small wind and hydro turbines, PV systems, or CHP plants. Some are connected to the 1-phase 230 V or 3-phase 400 V low voltage network but many to medium voltage circuits. Their principle of operation is to use the distribution network as a stable frequency and voltage reference and to inject energy back into the network whenever weather conditions allow or heat is needed from the CHP unit. The design of their control systems only allows operation when they are connected to a strong network and does not permit independent or islanded operation. The generators are rewarded for the annual energy, irrespective of when it is generated. As the distribution network was not designed for distributed generators, only a limited capacity of such generation can be connected to a circuit. The most common limitation to their connection is voltage rise caused by the reverse flow of power into a circuit that was only designed to supply loads.

The recent dramatic drop in the cost of PV generators has led to large numbers being connected to distribution networks. Small domestic PV systems are connected to the electricity circuits supplying each house and these reduce the amount of electricity imported from the grid. There is typically a ratio of more than 2:1 between the price of electricity paid by the domestic consumer and the price at which they can sell electricity back to the network. Hence it is desirable to consume as much energy as possible within the house but PV generation peaks at noon on sunny days when the domestic electrical load is often low. One solution is to store the surplus energy in batteries but these are expensive and the economics of their use to store PV generated electricity are at present marginal. An alternative approach is to supply loads such as water heating, washing machines, and dishwashers that can be operated whenever surplus energy from the PV system is available. Many modern appliances have controls to switch them on at a particular time and more sophisticated home smart control systems have monitors to detect surplus energy and switch the loads accordingly.

We can clearly see that all aspects of the electricity system are entering a period of great change. The way electricity is generated is determined increasingly by environmental concerns of climate change and air quality. The generation of electricity from renewable energy is increasing and traditional fossil fuelled generators are losing their base load function and are being used increasingly as reserve or back-up plant. The transmission system retains its pivotal role of managing system frequency but with more volatile generation and load. Smart meters and new monitoring equipment provide increased visibility of the network while HVdc allows long distance transmission of power over land and under the sea. The distribution networks are experiencing the most change with rapidly increasing distributed generation. If heat and transport are decarbonized using electricity as the means of supplying energy, then the distribution network will need greatly increased capacity.

Chapter 4
Nuclear power

Around 450 nuclear power plants, with a rated output of around 400 GW, produce approximately 11 per cent of the world's electricity. A further 60 reactors are under construction, mainly in Asia. Set against this increase in new generating capacity, older reactors are being decommissioned and so over the last fifteen years the amount of electricity generated from nuclear energy has remained broadly constant. Of all the sources of energy used for electricity generation, nuclear power is the most contentious with strong opinions both favouring and opposing its use. Some well-known environmentalists consider that the use of nuclear power is essential to limit climate change, while expressing reservations over its environmental impacts.

It is argued by those in favour of nuclear power that, when designed and operated correctly, nuclear generators are safe and have limited impact on the environment. Because of the very high energy density of nuclear reactions, the volumes of material that have to be extracted from the earth and transported and waste for disposal are much less than for fossil fuels. Nuclear power plants do not emit greenhouse gases when operating and generate electricity at constant output for months at a time. If access to a supply of fuel can be assured, the source of electricity is secure.

The opponents of nuclear power cite several reasons why the technology should not be pursued. The origin of nuclear power technology was in the development of weapons during the Cold War and there is continuing concern over the potential proliferation of nuclear weapons facilitated by the production of enriched uranium and plutonium. Recent experience in Europe has been that new nuclear power plants are expensive to build and subject to long construction delays. With their high capital but limited fuel costs, any delay in the commissioning of such capital-intensive generating plant greatly increases the cost of the electricity generated. The long-term disposal of high-level and intermediate-level nuclear waste remains unresolved in many countries. In addition there have been a number of accidents involving nuclear power reactors such as at Three Mile Island in 1979 and Chernobyl in 1986 and disruptive natural events such as the earthquake and tsunami in Fukushima in 2011. Although it can be argued that each accident had a different and specific cause of human error, or that a natural event was exceptional, they have eroded public trust in nuclear power.

There are two mechanisms by which nuclear energy could be used to create heat and so generate power; nuclear fission and fusion. The principle of nuclear fission is the splitting of heavy atoms such as uranium and is used in all existing electricity generating reactors. The chemical reactions of combustion (i.e. burning of fuel) result in the creation of new molecules but the number and type of atoms is unchanged. In contrast, nuclear fission splits the atoms of the fissile material and results in the creation of new and different atoms in a process known as transmutation. Compared with the chemical reactions of burning fossil fuels, very large amounts of energy are released in a nuclear reaction: 1 kg of natural uranium, when it is enriched and used in a reactor of the current common commercial design, releases energy equivalent to burning some 14 tonnes of coal.

The fusion of lighter atoms into heavier ones, in the same way that energy is released within the sun, offers very attractive possibilities for the generation of electricity. However, maintaining the conditions that are needed for fusion and extracting useful energy is technically extremely difficult. The generation of electricity from nuclear fusion remains the subject of continuing research and development but commercial operation of electricity generating fusion reactors is not anticipated for several decades.

Generation of electricity by nuclear power

Nuclear power reactors create heat, which is used to make steam that is then passed through a turbine to generate electricity. Once the heat is created, the steam cycle and generating equipment is similar to that of a coal fired power plant. A large nuclear reactor can be designed to generate up to 1600 MW_e, although such large individual generators can only be connected into strong power systems.

The heat energy is obtained from the fission reaction, which involves splitting the nuclei of uranium atoms by bombarding them with neutrons in a controlled chain reaction. The resulting fission products consist of new atoms and particles that together are lighter than the original uranium. Energy (E) and mass (m) are related through Einstein's famous equation $E = mc^2$. The speed of light (c) is very high and so very large quantities of energy are released by the small reduction in mass caused by the transmutation of a heavy uranium atom into two lighter ones.

The fission products are radioactive and so the spent fuel of a nuclear reactor requires careful shielding and disposal. The process of nuclear fission results in the creation of plutonium, an element that does not occur naturally. It is radioactive and is particularly hazardous if inhaled or ingested. Plutonium is used in the manufacture of nuclear weapons and some early reactors had the dual purpose of generating electricity and creating plutonium for military purposes.

Uranium is a naturally occurring element that is mined. Uranium metal extracted from the base ore consists mainly of two isotopes, uranium-238 (99.3 per cent by weight) and uranium-235 (0.7 per cent). Only uranium-235 is fissile and can take part in a chain reaction. When a neutron strikes an atom of uranium-235, multiple neutrons are released. If more free neutrons are created than are lost or absorbed within the mass of fissile material a chain reaction ensues.

In thermal reactors, a moderator slows down the neutrons to achieve more effective splitting of the uranium nuclei. Slow neutrons are more likely to cause fission than fast neutrons. The slow neutrons have similar kinetic energy to the target uranium atoms and so are said to be in thermal equilibrium—thus the nomenclature 'thermal reactor'. At the start of the fission process, such as when the reactor has been refuelled, neutrons are provided by start-up neutron sources that are distributed throughout the reactor core.

Fuels used in most electricity generating reactors have some component of uranium-235. A basic reactor consists of the fuel in the form of uranium oxide pellets packed into rods which is positioned in a moderator that slows down the neutrons. In current designs, the moderator is usually light or heavy water, or graphite. The great majority of modern reactors use light water both as the moderator and for heat transfer. Light water refers to conventional H_2O while heavy water describes deuterium oxide (D_2O). Deuterium is an isotope of hydrogen. Light water is less effective in slowing neutrons without absorbing them than heavy water and so light water reactors use enriched fuel with a higher fraction of fissile material.

Rods, of a material that absorbs neutrons (e.g. boron), are moved within the core to control the power of the reactor. The fission process in water reactors can also be controlled by adjusting the water conditions within the reactor or by changing the concentration of boric acid in the water.

A number of designs of reactor have been used with different coolants and types of fissile fuel. In Britain the first generation of nuclear power stations used Magnox reactors in which natural uranium in the form of metal rods was enclosed in magnesium-alloy cans. The fuel cans were placed in a core of graphite that acted as the moderator. This graphite core slowed down the neutrons to the correct range of velocities in order to provide the maximum number of collisions. Carbon dioxide gas was used to remove the heat to a steam generator.

Advanced Gas-cooled Reactors (AGRs) are still in use in Britain but are now coming towards the end of their service lives. Enriched uranium dioxide fuel in pellet form, encased in stainless steel cans, is used; a number of cans form a cylindrical fuel element, which is placed in a vertical channel in the graphite core. Carbon dioxide gas, at a higher pressure than in the Magnox type, removes the heat. The neutron absorbing control rods are made of boron steel.

In the USA and many other countries light water reactors are used. Light water reactors form the majority of the worldwide existing and planned fleet. These are either pressurized-water or boiling-water reactors. The ratio of pressurized-water reactors (PWRs) to boiling-water reactors (BWRs) throughout the world is around 60/40 per cent. Schematic diagrams of these reactors are shown in Figures 14 and 15.

In the pressurized-water type, water is pumped through the reactor and acts as both coolant and moderator; the water is heated to 345 °C at around 150–160 bar pressure. At this temperature and pressure the water leaves the reactor, at below boiling point, to a heat exchanger where a second hydraulic circuit

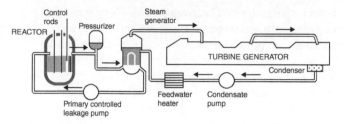

14. Schematic diagram of a pressurized-water reactor (PWR).

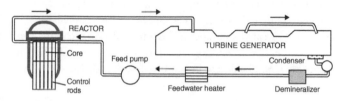

15. Schematic diagram of a boiling-water reactor (BWR).

feeds steam to the turbine. The fuel is in the form of pellets of uranium dioxide in bundles of zirconium alloy. The practical pressure limit for the pressurized-water reactor is about 160 bar, which limits its efficiency of electricity generation to about 30 per cent. However, the design is relatively straightforward and experience has shown this type of reactor to be stable and dependable. The compact design allows small PWRs to be used for the propulsion systems of nuclear powered ships and submarines.

The boiling-water reactor was developed slightly later. Inside the reactor, heat is transferred to boiling water at a pressure of 75 bar and a temperature of 285 °C. In the boiling-water reactor the efficiency of heat removal is improved by use of the latent heat of evaporation. The steam produced flows directly to the turbine. Boiling-water reactors are probably the cheapest to construct; however, the fuel is more complicated with different enrichment levels within each pin. The steam produced is saturated and requires wet-steam turbines. The steam and turbine are contaminated by radioactivity although with a short half-life.

The heavy water, CANDU type reactor was developed by Canada. Its operation and construction are similar to light water reactors but this design can use naturally occurring or slightly enriched uranium. Other reactor designs have been constructed including the RBMK type that was deployed in the former Soviet Union. This used naturally occurring uranium as a fuel, graphite as a moderator, and light water to transfer heat. The accident at Chernobyl revealed a number of flaws in its design including lack of comprehensive secondary containment and poor control at low output powers. The design is now obsolete outside Russia.

The nuclear fuel cycle

Uranium is widely distributed throughout the earth with significant concentrations in Australia, Canada, and Kazakhstan. Its extraction is only cost effective where there are relatively high concentrations (>0.1 per cent). Traditionally, the ore is extracted from surface or deep mines, crushed, and treated with solvents to separate the uranium oxides as yellowcake. Uranium can also be extracted by *in situ* leaching rather than conventional mining. In this process large quantities of water, injected with oxygen, are passed through the uranium deposits and the resulting liquid evaporated to leave yellowcake.

The uranium oxide can be converted directly into natural uranium fuel. The powdered uranium oxide is pressed into pellets that are heated to make a hard ceramic material. The pellets are then formed into tubes and assembled into fuel rods. More commonly the uranium is converted into uranium hexafluoride gas and passed through a gas centrifuge enrichment process that increases the concentration of fissile uranium-235. The gas is then converted into solid uranium dioxide powder and used to produce

Low Enriched Uranium (LEU) fuel. The resulting LEU fuel typically contains between 3 and 5 per cent fissile uranium-235 with the remainder being mainly uranium-238.

Some of the neutrons released during the fission process are absorbed by the uranium-238. This creates plutonium-239, which is itself a fissile material, and in some reactors fissioning of this plutonium accounts for as much as one-third of the energy generated. Plutonium-239 is a significant source of radioactivity and has a half-life of 24,000 years.

The simplest and most common fuel cycle is to use LEU fuel only once. After a period of operation of several years the spent fuel rods are removed from the reactor and stored for some years in deep water ponds. Radioactive decay of the fission products in the spent fuel creates considerable heat and the ponds provide cooling and a shield against radioactivity. After a period of storage on the reactor site, the spent fuel can be sent for permanent disposal.

If the facilities exist, the spent fuel can be reprocessed by chemically separating the fission products, the plutonium that has been created, and any remaining fissionable uranium. These are formed into mixed oxide (MOX) fuel and used in thermal reactors. MOX fuel is used in Europe and Japan and provides around 5 per cent of new nuclear fuel. Weapons grade plutonium or highly enriched uranium can also be disposed of in MOX fuel. The use of MOX fuel can require some technical modifications and re-licensing of the reactor. It is current practice to recycle reprocessed fuel only once.

Thermal nuclear reactors use a moderator to slow the neutrons to an energy level that enables effective fissioning of low enriched or naturally occurring uranium. An alternative approach is to use fast, high-energy neutrons to bombard highly enriched uranium, plutonium, or other fissile material. However the energy density of these fast neutron reactors is very high and they typically require liquid metal (e.g. sodium) cooling. Fast neutron reactors are technically demanding to construct and operate, and are not used for commercial electricity generation.

Some years ago, concerns over the availability of future supplies of uranium led to the construction of a number of prototype fast breeder reactors. In addition to heat, these reactors were designed to produce new fissile fuel by bombarding fertile nuclear material with neutrons. However the cost of fast breeder reactors together with the technical and environmental challenges of operating them led to most of these programmes being abandoned. It is now generally considered that supplies of uranium are adequate to provide nuclear fuel for the foreseeable future. This is particularly the case given the very large quantity of military grade fissile material in the world that is awaiting disposal.

Generations of nuclear reactors

It is conventional to discuss the development of nuclear reactors in terms of Generations I–IV. The first generation of reactors (Generation I) were designed in the 1950s and intended originally for the dual use of generating electricity and manufacturing plutonium for military purposes. These included Magnox reactors, the last of which was decommissioned in the UK in 2015, although the UK ceased production of military plutonium in the 1960s. Generation I reactors typically had an electrical output of less than ~400 MW_e.

Generation II began operating from the late 1960s and are the majority of reactors generating electricity at present. They include the common light water reactors (i.e. pressurized-water and boiling-water reactors based on US expertise) as well as the CANDU and AGR types and those developed by the former Soviet Union. They tend to have a higher output than earlier reactors (~1000 MW_e). After early problems and considerable development, many generating units have achieved reliable operation and capacity factors of more than 90 per cent.

Generation II reactors produce significant quantities of high and intermediate-level nuclear waste, particularly from the spent fuel. The USA has a large fleet of Generation II reactors and a policy of not reprocessing nuclear fuel in order to avoid separating plutonium and the proliferation of nuclear weapons. Reprocessing fuel, which is practised in the UK, France, and Japan, reduces the volume of intermediate-level waste and creates smaller amounts of high-level waste.

Generation III reactors are evolutionary developments of Generation II with particular emphasis on passive safety systems, increased thermal efficiency, and reduced cost of construction. With the benefit of operating experience, the lives of Generation II reactors have been extended to 50–60 years and Generation III units are intended to at least match this.

The large nuclear reactors of Generation III require very high investment to meet construction and financing costs. The size of the initial investment and the risks of delays and cost overruns make it very hard for private companies, operating in liberalized electricity markets, to take the commercial risk of such projects. Those schemes that are being developed are usually either government initiatives or are receiving considerable state support.

There is considerable interest in Small Modular Reactor units of ratings between 100 and ~300 MW$_e$. They would be based on established nuclear technology and made to a standard design in a factory. They might be located individually or in power parks. Small reactors were developed by the former Soviet Union to provide power for remote communities and there is a long experience of small PWR reactors for naval propulsion. It is hoped that by using established technology and building the reactors in a factory the construction times would be quicker and more predictable hence reducing the commercial risk of a project. The smaller project size and lower initial investment might make financing easier. However the cost of electricity generated by smaller reactors is likely to be higher than from larger units and the nuclear installation would still attract similar permitting and security costs, as well as possible public opposition.

Generation IV describes an initiative to develop radically different reactors either to generate electricity or to manufacture hydrogen. Several reactor designs are under investigation using thermal or fast neutrons and with liquid metal or gas for heat extraction. Most of these reactors are intended to breed nuclear fuel in a closed cycle but are not expected to enter commercial service for at least several decades.

Chapter 5
Renewable energy systems

Renewable energy systems use a broad range of technologies to capture and convert energy from natural energy sources that are not diminished as they are utilized. Most renewable energy comes from the sun. Natural processes convert heat from the sun into movements of wind and water in the atmosphere that are then used to generate electricity. Solar radiation provides the energy to grow biomass and sunlight is used to generate electricity directly through the photovoltaic effect. The heat generated by the sun's radiation is used to warm buildings and water.

Other than the sun, the only significant sources of renewable energy are the tides, which are caused by the gravitational attraction of the moon and sun combined with the earth's rotation, and geothermal heat from deep within the earth. However, these sources of energy are small compared to the power of the sun.

Renewable energy was used for thousands of years to provide heat, grind grain, and power ships but since the 18th century it has been progressively replaced by fossil fuels. It is now beginning to be used again instead of fossil fuels particularly for electricity generation.

There are many ways of harnessing renewable energy and methods of converting it into electricity, gas, or liquid fuel. Each has its advantages and disadvantages. Generally, the advantages of using distributed renewable energy sources are the limited, local environmental impacts. However, as the energy is concentrated (e.g. when water is collected into a large reservoir) the social and environmental impacts increase.

The disadvantages of renewable energy are that the resource is diffuse and its variable nature leads to a need for energy storage. Either the renewable energy resource can be stored, as in the case of dry biomass material, or the output of the energy conversion process can be accumulated, for example in large reservoirs of water, tanks of biogas, or electrical batteries. As bulk storage of electricity is not yet cost-effective, the current approach to dealing with variability of the resource is to integrate renewable generation into large power systems with generators powered by other sources of energy that have complementary power outputs and to use some fossil fuels when needed.

Table 7. Sources of world electricity in 2016

Electricity generation	Installed capacity GW_e	Electricity generated TWh	Electricity generated %	Annual capacity factor %
All	6,508	23,766	100.0	42
Fossil	4,017	15,367	64.7	44
Hydro	1,092	4,001	16.8	42
Nuclear	353	2,469	10.4	80
Renewables excluding hydro	893	1,967	8.1	25
Wind	468	952	4.0	23
Solar	297	343	1.4	13
Biomass and waste	114	567	2.4	57
Geothermal	12	77	0.3	73

Table created using data from US Environmental Information Administration, EIA Beta Energy Statistics (*Sep 2018*).

The success of any renewable energy project depends partly on the technology but relies entirely on the resource and how the scheme is integrated into the environment. Without a good energy resource no renewable energy scheme will be cost-effective. The low energy density of renewable energy means large pieces of equipment and areas of land are needed to capture energy and convert it into useful quantities of power. As with any energy project, if the proposal does not address its environmental impact adequately and carry public support then the civic authorities will not allow it to be built.

Table 7 shows the fuels at present used for electricity generation throughout the world. Almost 25 per cent of electrical energy is generated from renewables including hydro. The installed

Box 6 Capacity factor

The annual capacity factor is the ratio of electrical energy generated over a year (8760 hours) to that if the generator had operated continuously at full output. It is calculated by:

$$\% \, annual \, capacity \, factor = \frac{Annual \, electricty \, generated \,\, [TWh] \times 100}{Installed \, capacity \,\, [TWe] \times 8760}$$

The reasons why a generator might not operate include:

- Lack of renewable resource or fuel
- Technical breakdowns and maintenance
- Lack of market for the electricity

The capacity factor shows how effectively the generator is being utilized. It highlights a common theme of this book that it is essential to distinguish between power (installed capacity) and energy (electricity generated). Renewable energy sources that are powered directly from the sun or wind have low capacity factors and so a large capacity of generating plant is needed.

capacity of wind and solar photovoltaic generation is increasing rapidly but from a relatively low base. The annual capacity factor (see Box 6) varies from around 80 per cent for nuclear generation which is intended to operate almost continuously at full output to 13 per cent for photovoltaic systems that are powered directly by the sun.

Solar energy

The sun is a spherical collection of hot gases with an internal temperature of up to 20 million °C. Its centre is a nuclear fusion reactor converting hydrogen into helium and heavier elements, so releasing energy. Energy generated by the fusion reaction at the centre of the sun is transmitted through its hot gases to the surface, which is at a temperature of around 5800 °C. Radiation is emitted from the surface of the sun and travels through space to arrive at the outer surface of the earth's atmosphere. The intensity of the sun's radiation at the outer surface of the earth's atmosphere (known as the solar constant) is around 1367 W/m².

Once the radiation enters the earth's atmosphere it is reflected and scattered by clouds and various gases, particles, and vapours. These interactions reduce the intensity of the radiation and deflect some of it from its direct path from the sun. At sea level on the earth's surface, the intensity of irradiance drops to less than 1000 W/m² with its spectrum modified by gases and aerosols in the atmosphere which selectively absorb certain wavelengths.

The solar resource reaching the earth's surface can be considered in two parts. Direct radiation is received from the sun without it having been scattered by the atmosphere. Diffuse radiation is received from the sun after its direction has been changed through atmospheric scattering. Windows and flat plate solar collectors make use of both direct and diffuse radiation while concentrating collectors only focus direct radiation. In temperate latitudes much of the radiation received is diffuse so making concentrating solar collectors ineffective.

The unit used to describe the power of the solar resource is irradiance (W/m²). This is the rate at which solar energy strikes a surface. In contrast, insolation is the incident energy per unit area on a surface and is found by integrating the irradiance over a specified time, typically a day. Its units are J/m² or kWh/m². Some international standards use the term irradiation instead of insolation to describe incident solar energy.

Near the equator, the daily solar insolation is fairly constant over the year, only varying with cloud and local atmospheric conditions. However, in temperate latitudes there can be a great difference between the insolation in summer and winter, and hence the energy available. Figure 16 shows the total solar irradiance measured on a horizontal surface in California over five days of January. The daily variation of irradiance from sunrise to sunset due to the position of the sun in the sky as well as the effect of cloud cover can be seen. In this week, the irradiance peaked at just less than 600 W/m² at midday while the daily insolation was modest, ranging from 1.5 to 2.9 kWh/m² over these days of a winter month.

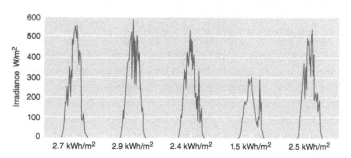

16. **Solar irradiance on a horizontal surface in California over five days of January. The daily insolation is given for each day below each daily irradiance curve.**

Photovoltaic (PV) electricity generation

Generating electricity directly from solar radiation became possible in the 1950s through the development of solid-state semiconductor electronics. It is now the fastest growing source of electricity generating capacity with a worldwide annual rate of increase of capacity of up to 30 per cent/year. This high rate of growth is partly due to the financial support offered by a number of governments for PV generation but also because in recent years there has been a dramatic reduction in the price of photovoltaic modules as manufacturing volumes have increased.

Photovoltaic modules have no moving parts and are extremely robust. They can operate for more than twenty years with minimal maintenance and their installation has limited environmental impact. The solar panels are mounted either on the roofs of buildings or are supported on ground mounted structures in large solar farms. Some innovative buildings have photovoltaic modules integrated into their façades or roof. The majority of solar panels that are currently installed export their power to the electricity grid through a dc to ac converter. These converters are designed to operate only when there is a stable grid supply and so do not function during power cuts.

In addition to grid-connected systems, off-grid photovoltaic systems with batteries can supply small amounts of electrical power in remote areas that do not have a grid electricity supply. The electrical energy produced is expensive but its use can be justified by the value of the load. For example, off-grid PV systems are commonly used for lighting and for vaccine fridges in remote areas of developing countries.

The majority of current commercial photovoltaic systems use crystalline silicon solar cells and their performance is stable with little reduction in output over time. A diagram of a silicon

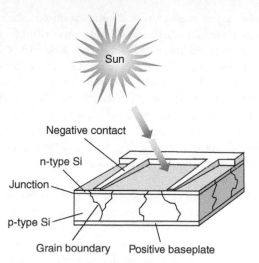

17. Schematic of a poly-crystalline silicon solar cell.

photovoltaic cell is shown in Figure 17. Crystalline silicon is used to form a positive–negative (p–n) junction that is illuminated by sunlight. Pure silicon is either grown into a single crystal (mono-crystalline) or cast in ingots making a number of large crystals (poly-crystalline). The silicon is doped with boron to create the p-type semiconductor as the crystalline silicon is manufactured. The p-type silicon is sliced into wafers some 100–200 μm thick using a diamond tipped saw. Then, in a separate process, phosphorus is diffused into the upper surface of the wafer to form a thinner layer (~0.3 μm) of n-type semi-conductor. The p–n junction is formed where the two semi-conductors meet. Electrical contacts are applied to the upper and lower surfaces of the photovoltaic cell and an anti-reflection coating is placed on the upper surface.

In a photovoltaic cell, the light photons excite electrons that are pulled across the cell by the electric field created at the p–n junction of the two dissimilar electronic materials. The movement of the electrons creates a current. Crystalline silicon solar cells have been the subject of an intense development effort for more than thirty years and now have reached cell efficiencies of up to 26 per cent for test cells in the laboratory and up to 21 per cent for commercially available complete modules. Although silicon (sand) is common in the world, the manufacturing process to purify the bulk silicon material is expensive and takes a considerable amount of energy. Depending on the solar resource where it is installed, a crystalline silicon solar module may take up to five years to generate the energy used in its manufacture.

Many other materials can be used to create solar cells and there is very active research and development to create so-called second and third generation solar cells. These use smaller quantities of photovoltaic material and can be processed with less energy. However these cells can create greater environmental impact and their performance can degrade over time. So far none of these new photovoltaic technologies challenges the commercial dominance of first generation crystalline silicon cells.

An individual photovoltaic cell produces only a few Watts of dc electrical power and crystalline silicon wafers are fragile. Hence the cells are collected into modules to give more useful levels of power and packaged to provide mechanical and environmental protection. A typical solar module might produce 200 Watts at around 30 Volts dc in bright sunlight.

The peak power output of a photovoltaic system is specified (and tested) under Standard Test Conditions (STC) of irradiance of 1000 W/m^2. As shown in Figure 16 these conditions rarely occur in practice. For a first estimate of the energy output of a solar PV system the rated power output at STC (defined by the manufacturer) is multiplied by the insolation experienced at the site (obtained from a database or site measurements). In the UK the daily solar insolation resource can vary between 5 kWh/m^2 in summer and less than 1 kWh/m^2 in winter. Thus a 4 kW PV system might generate 20 kWh of electricity over a summer day but only 4 kWh or even less in a winter day. If a 4 kW generator were supplied from a stored fossil fuel, such as gas, and operated continuously then it would generate $4 \times 24 = 96$ kWh each day. However, because of the limited solar resource, a PV system in the UK would produce only a fraction of its full output, that is, a daily capacity factor of 4/96 or 4 per cent in winter and 20/96 or 21 per cent in summer. This simple calculation illustrates a key characteristic of renewable energy schemes that although the peak power may be large the energy generated depends entirely on the strength of the renewable energy resource.

Solar thermal energy

Solar cells capture the energy in light from the sun to generate electricity directly using the photovoltaic effect. In contrast, solar thermal energy systems use the sun's energy to heat buildings or a fluid, often water. Many solar thermal systems naturally include a simple energy store such as heat in a hot water tank or in the fabric of a building.

Solar thermal energy systems may be divided into those processes that use a low temperature (less than ~150 °C) and a high temperature (greater than ~150 °C). Although the use of low temperature solar heat appears simple it is of great practical importance. Much of the fossil fuel burnt in countries with temperate climates is used in heating buildings and providing hot water. Replacement of even a fraction of this by solar energy offers significant financial and environmental benefits.

Glass is an essential component of many solar thermal systems. It has the important property of allowing radiation from the sun to pass through it while trapping warmth behind it. The use of glass to exploit solar thermal energy to heat buildings was known to the Romans and remains an important aspect of building design. For maximum solar gain a building should be orientated east–west with large windows located on the south side (if it is in the northern hemisphere). Any windows on the north side, which receive little irradiance, are kept small to reduce heat loss. Solar radiation enters the building through the large, south-facing double- or triple-glazed windows and warms the internal structure of the building. The glass windows then trap this warmth. Buildings that use this passive solar gain are usually designed to store heat so that they maintain a reasonably constant temperature.

Careful design is necessary to make use of passive solar gain. In the winter when heating is required, the solar resource is limited so large windows are needed. In the summer the need for heating is small but the large windows can then cause the building to overheat. Measures to avoid overheating include the use of manually operated or automatic blinds to reduce the irradiance entering the building, and overhangs positioned above the windows which allow irradiance to enter the building in winter when the sun is low in the sky, but block the irradiance in summer when the sun is high in the sky. External shutters, as used in many hot climates, are particularly effective in preventing too much radiation from entering the building.

After space heating, the supply of hot water consumes the greatest proportion of heat energy in buildings. Solar water heating systems are common and even mandated by regulations in some countries with a good solar resource. An unglazed flat plate collector can only provide an increase in water temperature of up to 10 °C above ambient. Glazed flat plate collectors use one or more sheets of glass to increase the water temperature up to 50 °C above ambient. Temperatures of more than 50 °C above ambient can be obtained with evacuated tube collectors.

High temperature solar thermal systems to generate mechanical power were first demonstrated more than 100 years ago but their use for electricity generation requires a good solar resource of direct irradiance and the technology is expensive. Recently the number of prototypes and demonstration plants of large-scale high temperature solar thermal systems for electricity generation using parabolic troughs or solar towers and heliostats (mirrors that track the sun) has increased. This technology is known as Concentrated Solar Power (CSP) and is being actively developed for areas of the world with a good solar resource, such as deserts, but has yet to be used on the same scale as photovoltaic generation.

Most of the CSP plants that are in commercial operation use line-focusing parabolic troughs to concentrate the irradiance (Figure 18). Parabolic troughs of silver coated glass mirrors are placed along an axis, usually north–south, and the troughs are rotated through the day by an actuator and control system to track the sun from east to west. Each trough assembly of mirrors is up to 150 m long, typically in 15 m sections. Each section is made up of curved mirrors with the cross-section of a parabola supported on a space frame to form a rigid rotating structure.

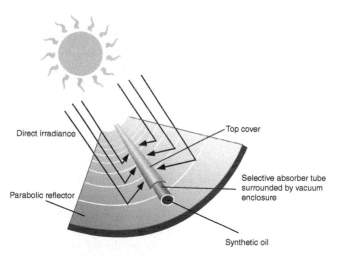

18. **Parabolic trough solar collector.**

The parabolic trough focuses the direct irradiance of the sun on to an absorber pipe that is located at the focus of the parabola. Heat is extracted from the absorber pipes by a flow of synthetic oil and the hot oil is collected at a central generating plant. Then, through a heat exchanger, the heat is used to power a steam turbine. It is common for CSP plants to consist of a number of trough assemblies to give an electrical output power rating of up to 200 MW$_e$ using steam at a temperature of up to 400 °C. An advantage of CSP is that a thermal heat store and even auxiliary gas boilers can be included in the system to allow generation throughout the night.

Hydropower

Water wheels and then turbines were used to drive the machinery of many factories of the early industrial revolution and hydropower was the energy source for some of the earliest electrical generators. Hydropower at present supplies around 17 per cent of electrical energy worldwide. However, in developed countries most of the sites that are suitable for hydro power stations have been utilized and new large schemes are likely to have significant environmental and social impacts and so face intense public scrutiny. In less populated parts of the world there remain sites suitable for hydropower schemes that have yet to be developed, but further schemes with large reservoirs will be controversial.

The advantages of hydropower for generating electricity are that its operating costs are very low and many hydro power stations are controlled remotely with few staff permanently on site. The power output of the turbine generators can be adjusted within seconds and if a reservoir is used then water can be stored and electricity generated when required. Thus hydropower is very useful when operating an electrical power system. It is often used to balance the load demand, which varies throughout the day, and to provide reserve capacity in case of faults. The civil works of a hydro scheme can have a lifetime of up to 100 years.

There are disadvantages; first the high capital costs, particularly of the civil works. Also, large reservoir hydropower schemes can have a significant environmental and social impact, displacing rural people from their land and destroying important natural habitats. The power output of a run-of-river scheme varies depending on the rainfall and the vegetation and soil of the catchment area. Such a scheme, without energy storage, may be fully utilized for as little as 20–30 per cent of the time. This level of utilization (or capacity factor) is similar to that of a wind farm. When large reservoirs are first flooded, decomposing submerged vegetation can lead to significant emissions of methane (a powerful greenhouse gas). The failure of a large dam can lead to great loss of life.

Large hydropower schemes that have been developed include the 22 GW Three Gorges scheme in China and the 14 GW Itaipu scheme on the border of Brazil and Paraguay. However, around 10 per cent of the worldwide capacity of hydropower is from a large number of small schemes (less than 5–10 MW). Smaller hydro schemes, particularly those without large reservoirs, can be arranged to have fewer social and environmental consequences and are being actively encouraged in many countries.

The energy of a hydropower scheme comes from rainfall driven by the hydrological cycle. Solar radiation causes evaporation from land and sea, clouds form, and rainfall results. Some of this precipitation is lost through evaporation, transpiration by vegetation, and surface absorption. The remaining water flows into streams and then rivers. The catchment areas of land act to concentrate the precipitation into a useful flow of water and, depending on their characteristics, will store water and result in more or less uniform flows in the rivers.

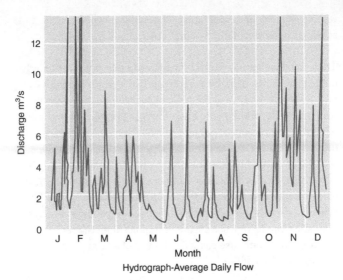

19. Average Daily Flow of a river with variable discharge over the year.

The flows in a river can be described by a time series of daily flows (known as a hydrograph). Figure 19 shows a hydrograph of a river in Wales. The soil of the steep catchment is rocky with little vegetation. It may be seen that the daily flows are very variable. Unless a reservoir is included in the scheme, the output power of a hydro-generator will vary with rainfall. This variable output is typical of small hydro schemes that have small catchments or limited water storage provided by the soil or vegetation.

The power that can be generated from a hydropower scheme depends on the flow of water and the height through which the water falls. The efficiency of large turbines can approach 95 per cent. The schemes can be run-of-river, generating in proportion to the flow of the river, or fed by a reservoir which stores energy to be used when needed. Pumped storage schemes have an upper and lower reservoir. Water falling from the upper reservoir generates power when it is needed and the same turbines are used to pump it back when electricity is cheap.

Hydro schemes are classified in a number of ways including by the height through which the water falls (known as the head). High head schemes have a useful height of greater than around 200–300 m, medium head between 30 and 200 m, and low head less than 30 m.

Figure 20 is a simple schematic of a high head hydro scheme. The penstock is a pipe able to withstand high pressures that takes water from a high level intake to a turbine positioned at a lower level. The flow of water under pressure from the height operates the turbine to which a generator is coupled either directly or, for smaller generators, through a gearbox. After passing through the turbine, the water exits into the tail waters and rejoins the river system. A surge tank limits the dynamic pressures that can arise if the flow of water is suddenly restricted. For some types of turbine a draft tube acts to increase the effective head of the scheme.

Medium head schemes often have the turbines and generators in a powerhouse at the base of the dam with a limited length of penstock and use a large flow rate to generate considerable

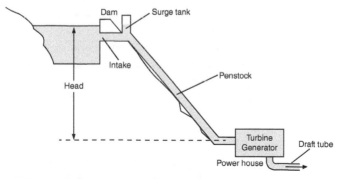

20. **Schematic diagram of a high head hydro scheme.**

amounts of power. Some of the largest hydropower schemes use this arrangement.

Low head schemes have smaller power outputs and usually do not include large reservoirs. They are often located along waterways and the civil works will then include locks for boats and barges and a fish ladder for migrating fish. The draft tube is a particularly important element of a low head scheme as it can significantly increase the head available.

Wind power

Horizontal axis windmills were important for grinding corn throughout Europe from the 12th century, while vertical axis windmills were used even earlier in parts of Iran and China. In the USA, small multi-bladed wind turbines became widespread for pumping water from around 1850 and remained important until rural electrification in the 1930s allowed electric pumps to be used more cheaply and conveniently. There were experiments and demonstrations of using wind energy to generate electricity throughout the 20th century. However it is not easy to construct a reliable and cost-effective electricity generating wind turbine and for most of that century the low price of oil encouraged fossil fuel use. This changed when the high oil prices of the early 1970s stimulated major government programmes of research and development into large wind turbines. In a parallel initiative, subsidies and tax breaks supported the installation of smaller turbines in Denmark, Germany, and California. A period of some forty years of development followed and wind energy is now an established and increasingly important source of electricity. Figure 21 shows the remarkable increase in size and output of wind turbines over the last twenty-five years.

Each wind turbine is a significant source of generation. Offshore turbines can be up to 12 MW rating and have rotors up to 200 m in diameter. Turbines for onshore sites tend to be smaller, typically

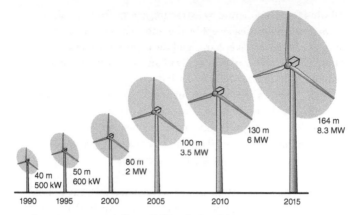

40 m
500 kW
1990

50 m
600 kW
1995

80 m
2 MW
2000

100 m
3.5 MW
2005

130 m
6 MW
2010

164 m
8.3 MW
2015

21. Largest commercially available wind turbines.

up to 3 MW and 100 m rotor diameter, both to reduce visual intrusion and for ease of transport on rural roads. The onshore turbines are either sited individually or grouped into wind farms, often of capacities between 10 and 50 MW. However larger wind farms of around 500 MW capacity have been built in sparsely populated areas of western China and on the Great Plains of the USA, or offshore in the shallow water of the North Sea. Once planning (permitting) permission is obtained an onshore wind farm can be constructed quickly, within 6–12 months. On high wind speed onshore sites the cost of generation is comparable to that from fossil fuels.

The main disadvantages of wind power are the visual impact of the turbines, noise from the blades, and the variable nature of the wind and hence the output power. In some countries there is increasing development of wind farms offshore where the wind is smoother with higher average speeds than on land. Also there is less visual impact and noise is less important. However the increased capital and operating costs in the demanding offshore environment give a higher cost of the electrical energy generated.

All wind turbines operate by extracting energy from the air that flows through the area swept by the rotor. The power generated is proportional to the swept area and the cube of the wind speed. These simple relationships apply to any wind turbine and explain the large size of the rotors and why every effort is made to place the turbines on sites with high wind speeds. Unlike hydro turbines, which are located in ducts and can have very high efficiencies, wind turbines operate in an open flow and have a maximum theoretical efficiency of 59 per cent. In practice, efficiencies of over 45 per cent are achievable.

The rotor of a wind turbine can be constructed with any number of blades rotating about either a horizontal or vertical axis but it is now accepted practice to use three-bladed, horizontal axis, upwind rotors as shown in Figure 22. The maximum speed of the blade tips is limited to control noise and so large turbines rotate slowly. A conventional electrical generator operates at a much higher rotational speed than the aerodynamic rotor, so the blades are coupled to the generator through a speed-increasing gearbox. The entire nacelle rotates on a bearing at the top of the tower, to always point into the wind.

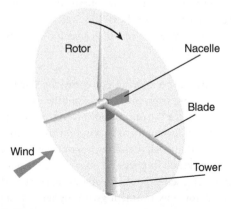

22. Wind turbine.

This architecture was only established after a period of experimentation in the 1970s and 1980s. During that time two-bladed and even single-bladed horizontal axis turbines were developed and offered commercially. Three-bladed wind turbines are visually more pleasing than two- or single-bladed designs and so became dominant. The importance of visual appearance and the public perception of a wind farm project (or any other renewable energy scheme) cannot be overstated. In the future, lower cost, two-bladed designs may be used for very large offshore wind turbines where visual impact and noise are less important.

Vertical axis wind turbines, using either straight vertical blades or forming an egg-beater shape (Darrieus type), were also developed and sold during that time of experimentation. However they have not become commercially competitive. In a vertical axis wind turbine, there is a large cyclic torque on the rotor caused by the blades experiencing different apparent wind speeds as they move upwind and downwind. These cyclic forces have to be resisted by a strong, and hence expensive, rotor and support structure.

All modern electricity generating wind turbines use the lift force (as in an aircraft wing) to drive the rotor. The blades rotate and occupy only a small fraction of the area swept by the whole rotor. The rotor then acts as a concentrator of the energy in the wind, so the energy generated over a wind turbine's life is much more than that used for its manufacture and installation. An energy balance analysis of a 3 MW wind turbine showed that the expected average time to generate a similar quantity of energy to that used for its manufacture, installation, operation, transport, dismantling, and disposal was 6–7 months. A similar time was calculated for installation both onshore and offshore as the higher wind speeds offshore balanced the increased energy used during installation.

The wind at any site varies with the weather and local conditions. Figure 23 shows the output of a small wind farm over a summer

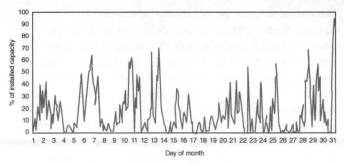

23. Output power of a small wind farm.

month of modest wind speeds. The variation in wind speed and hence power is caused by a combination of changes in the weather over a day or two and higher frequency turbulence created by the roughness of the ground over which the wind flows. The variations in wind speed also lead to changes in the forces on the blades and components of a turbine. The forces on a wind turbine are proportional to the square of the wind speed and in the early days of wind turbine development, when the effect of the variations in wind speed was not fully understood, many of the early prototypes failed. Gearboxes were particularly vulnerable. Some manufacturers have developed wind turbines without gearboxes but these need large diameter electrical generators to operate at the same rotational speed as the aerodynamic rotor. The choice of whether a gearbox or large diameter generator is used is a matter of preference of the manufacturer.

All modern wind turbines operate at variable rotational speed. This is partly to reduce mechanical loads on the turbine caused by the turbulence in the wind but also to achieve maximum efficiency over a range of wind speeds. Variable speed operation is achieved by connecting the generator of the wind turbine to the power network through power electronic converters.

Until around the year 2000, the installed generating capacity of wind turbines was so low that its output was viewed by the operators of the electricity system as negative load that supplied energy but played no part in supporting the operation of the power system and maintaining its stability. Since then, with the greatly increased capacity of wind generation, turbines are required to contribute to the operation of the power system particularly during and after network faults and disturbances.

In an onshore wind farm project, the turbines and their erection will account for around 65–70 per cent of the total cost with the remainder for foundations, site roads and lay-down/crane areas, electrical systems, and project development costs. Offshore the increased cost of foundations, submarine cables, and access reduces the turbine fraction to around half of the much higher project costs. Commercial developers of wind farms often prefer larger projects as the fixed costs, particularly grid connection and project set-up, management, and financing costs, may be spread over a bigger investment. However there are individuals and community groups who develop smaller wind farms and single turbine projects. The advantage of community involvement in the project is that planning permission is more readily obtained if it is seen that there is tangible benefit to local people.

Large wind farms extend over wide areas and the individual turbines can be more than 150 metres high. Before permission is given for construction a comprehensive environmental impact assessment is undertaken. In many countries environmental restrictions are a major cause of projects not proceeding. The most important environmental constraints include avoidance of National Parks or other areas designated as being of particular amenity or scientific value. It is also necessary to ensure that no turbine is located so close to dwellings that a nuisance will be caused by noise, visual domination, or light shadow flicker.

The visual appearance of the wind farm particularly from important public viewpoints is considered. If, within the wind farm perimeter, there are areas of particular ecological value, due to flora or fauna, then these need to be avoided as well as any locations of particular archaeological or historical interest. Communication systems such as microwave links, TV, radio, and particularly aviation radar may be adversely affected by wind turbines and these need early consideration.

Bioenergy

Biomass supplies some 10 per cent of the world's primary energy, mostly from traditional fuels in developing countries. In these rural societies wood, charcoal, crop residues, and dried animal dung are widely used for cooking and heating. It is estimated that 3 billion people worldwide use biomass for cooking. In the industrialized world, the use of bioenergy to generate electricity and as fuel for road vehicles is growing rapidly although from a relatively low base.

Bioenergy has a number of very useful attributes. Biomass can be stored as a dry solid or converted into a gaseous or liquid biofuel. Hence although bioenergy is a concentrated form of solar energy it does not depend on the instantaneous irradiance of the sun and can be stored and used when needed. The other renewable energy technologies generate electricity with very low emissions of greenhouse gas once the equipment is manufactured and installed. When biomass is burnt and converted into useful energy the CO_2 that was absorbed from the atmosphere during photosynthesis is released. In practice some fossil energy is needed for the cultivation, fertilizing, harvesting, transport, and processing of biomass and so careful analysis is required to determine the lifetime environmental costs and benefits of any biomass scheme. When land is cleared to grow biomass crops, CO_2 that has been stored in the soil can be released into the atmosphere.

Biomass requires a considerable area of land, and its energy density by volume and weight is low in comparison to fossil fuels. Thus its costs of collection and transportation are high and this limits the practical size of the majority of biomass or waste fuelled electricity generating plants to less than ~20 MW$_e$. The exceptions to this are some very large electricity generating units in Great Britain (up to 500 MW$_e$) that are fuelled by wood pellets imported from the USA.

Table 8 lists the main sources of bioenergy. It may be seen that these sources are very varied and a range of processing techniques are required to transform the biomass into useful energy.

The techniques for processing biomass into bioenergy can be divided into thermochemical and biochemical processes as well as the mechanical extraction of oil from plants. Thermochemical

Table 8. Main sources of bioenergy

Main sources of bioenergy	Details
Food (and fodder) crops	Edible parts of sugar, starch, and oil plants that are traditionally grown for food for humans or animal fodder. Food crops at present being used for biofuels include wheat, maize, soya, palm oil, oilseed rape, and sugar cane.
Agricultural residues	By-products from crops such as wheat straw, rice husks, and sugar cane residue (bagasse). Also slurry and manure from animals.
Forestry and forest residues	Woody material from existing forests (which may be managed) plus residues from sawmills and tree pruning.
Municipal waste	Food and domestic waste, sewage, and other biological waste.
Energy crops	Fast growing trees and grasses grown for energy e.g. willow and miscanthus, oil crops such as jatropha.

processes use heat and catalysts to transform biomass into useful energy by combustion or gasification. Biochemical processes use enzymes and micro-organisms in alcoholic fermentation or anaerobic digestion. Vegetable oils are mechanically extracted from plant seeds, processed, and either used directly in compression ignition engines or converted into biodiesel.

Solid biofuels are burnt in devices that range in size and sophistication from small rural household cooking stoves in developing countries to steam raising boilers for large electricity generators. In its natural state solid biomass is not easily flammable and needs to be dried to reduce the water content before use. It then follows a complex thermochemical conversion process in order to burn satisfactorily.

In developing countries, traditional biomass is often used for cooking over open fires. This is very inefficient and damaging to health; it is estimated that the fumes from open fires lead to 4 million excess deaths worldwide annually. Both the efficiency and impact of burning traditional biomass can be improved dramatically if simple stoves are used, but their widespread introduction has been slow for complex social and cultural reasons. The overall efficiency of electricity generation by combusting biomass is relatively low, being only 15–20 per cent for small plants rising to 30–35 per cent in large units. Both the control of emissions and the disposal of waste from biomass combustion are more difficult than for more homogeneous fossil fuels.

Unprocessed biomass (e.g. straw, crop residues, and wood) has a relatively low energy density compared to fossil fuels and so is expensive to handle. It has to be stored carefully in dry conditions. The gasification of biomass changes a heterogeneous organic material into a homogeneous gaseous product that can be cleaned of impurities and then stored and transported more easily. Gas is a much more convenient fuel than solid biomass and can be burnt at higher temperatures thereby increasing the thermodynamic efficiency of its use. Gas can also be compressed and pumped.

Biomass can be converted into gas by subjecting it to a high temperature with limited air or in the absence of oxygen. Considerable heat energy is required for gasification and commonly the heat is provided by partial oxidation, that is, by burning some of the biomass. Alternatively in an indirectly heated gasifier the heat is supplied from an external source. The flammable products of thermal gasification depend on the feedstock and process but typically are a mixture of carbon monoxide, hydrogen, and methane. The term syngas is commonly used to describe the product of all forms of thermochemical gasification.

Anaerobic digestion is a natural biochemical process in which biomass is broken down by micro-organisms in the absence of oxygen. Naturally occurring micro-organisms digest the biomass and form biogas, whose main flammable constituent is methane, and a solid/liquid residue. The biogas can be burnt for cooking, lighting, and heating or used to power internal combustion engines to generate electricity. The methane can also be cleaned and injected into the natural gas supply grid. The residual solid matter from anaerobic digestion is known as digestate and, depending on the feedstock, can be a valuable fertilizer.

Biomass is converted into liquid fuel for vehicle engines in two quite different ways. Alcoholic fermentation is used to produce ethanol from sugar or starch crops. The ethanol is used as a partial or complete substitute for gasoline in spark ignition engines. Alternatively oil is extracted by pressing the seeds of crops such as rape, sunflowers, or olives and either used directly in compression ignition engines or processed by transesterification into biodiesel.

As with all energy sources the use of bioenergy has both environmental and social impacts that must be managed. The emissions and waste products from biomass processing need careful management and the environmental impact of large areas of monoculture energy crops has been questioned.

Many of the existing bioenergy processes, particularly for the manufacture of vehicle fuels, use sugar, starch, or oil crops that could otherwise be used for food. The use of food crops as a source of bioenergy and the consequential effect on world food prices is controversial. Current research and trials are investigating how non-food crops can be grown on poor quality land that is unsuitable for other uses and converted into biofuel.

It is anticipated that in the future it will become possible to capture CO_2 emitted from electricity generating units or large industrial plants and sequester it underground. If applied to installations using biofuels, this could result in the bioenergy cycle reducing CO_2 in the atmosphere (i.e. overall negative CO_2 emissions). However, this technology, of Carbon Capture and Storage (CCS), has yet to be demonstrated to be cost-effective even with fossil fuels let alone more demanding heterogeneous biomass.

Geothermal energy

Geothermal energy is the energy stored as heat beneath the earth's surface. Worldwide there is at present around 12 GW of electricity generating capacity powered by geothermal energy with a mean capacity factor of 72 per cent. There is some 3.5 GW of generating capacity in the western USA and significant developments in the Philippines and Indonesia. In Iceland geothermal energy is used to generate around 13 per cent of the electricity from 600 MW_e of capacity and it meets almost all the space heating needs of the country. In France more than 350 MW_{th} of heat at 60–80 °C is extracted from the Paris sedimentary basin for district heating.

In many schemes, the geothermal resource is not strictly renewable as the heat is extracted at a rate greater than that at which the reservoir is recharged and so the output declines over a number of years. However, geothermal energy shares the characteristics of other renewable energy schemes of creating very low CO_2 emissions during operation and relying critically on the local energy resource.

The earth has an inner core of solid iron believed to be at more than 5000 °C. It is thought that the heat was created partly as the earth was formed from dust and hot gases and also later from radioactive decay and the heat of crystallization of molten rock. The core is surrounded by successive layers, including one of hot molten rock or magma at temperatures of up to 1000 °C on which the earth's crust floats. Heat is emitted from the earth's surface at an average rate of around 0.1 W/m². The average temperature gradient near the surface of most of the earth is between 25 and 30 °C per kilometre of depth. However in volcanic areas, where the continental plates meet, much higher temperature gradients can be found as the magma is closer to the surface. Electricity generation from geothermal energy is only cost-effective in volcanic regions with high temperature gradients as the cost of drilling wells for the extraction of the heat dominates the economics of any scheme.

Geothermal energy can be used in several ways. Near the surface of the earth at depths of only 10–15 metres the temperature of the ground is essentially constant throughout the year, at around the mean of the air temperature. This constant temperature can be used by geothermal or ground source heat pumps as the sink or source of heat for either heating or cooling. Depending on the details of the application, electrically powered heat pumps can have coefficients of performance (the ratio of electrical energy used to heat transferred) of 3–4 and so their use is an effective way of using low carbon electricity generated from nuclear or renewable energy to provide heat or cooling.

In certain parts of the world an increased heat gradient and local heat load makes it economic to utilize heat directly from aquifers or old mine workings at depths of up to several kilometres. Such schemes require a highly permeable reservoir that allows large volumes of water to be extracted with reasonable pumping costs.

In volcanic regions where there is a large heat gradient, heat energy for electricity generation is extracted from wells that are usually several kilometres deep. A few schemes use dry steam directly from a geothermal reservoir to turn conventional turbines. More commonly high-pressure hot water changes into wet steam as it comes to the surface and the steam and hot water are then separated. The steam is used in a conventional turbine. Binary cycle power plants use a heat exchanger to transfer the heat to a second loop of fluid with a lower boiling point that is then used to power a turbine.

Marine energy

The oceans cover more than 70 per cent of the earth's surface and energy from the tides and waves is an important potential source of electrical power. However, the marine environment is extremely demanding not least because of the wide range of forces that any marine energy conversion device will experience. In normal operating conditions a marine device must extract energy effectively from relatively small and benign movements of water caused by the waves and tides but it must also survive damaging extreme storms. In addition, installing and gaining access to an offshore marine energy device for maintenance is difficult, while transmitting electrical power to the shore using submarine cables is expensive. Seawater is corrosive and so any materials used must be carefully selected and protected. Hence although there has been considerable research and development effort over many years and a number of demonstration projects have been installed, marine energy has yet to achieve a commercial breakthrough and there is at present less than 1 GW of marine energy electricity generating capacity in service worldwide.

Marine energy describes three distinct technologies. These are generation by: tidal range, tidal stream, and wave power.

The tides are caused by a very low frequency water wave that is created by the gravitational attraction of the moon and the sun and the rotation of the earth. This wave with a period of around 12½ hours leads to variations in the height of the water and the creation of marine currents as it approaches the shoreline. In contrast, ocean waves are created by winds, which are ultimately due to the heating of the earth by solar energy and contain a range of higher frequencies. Both tidal range and tidal stream generation have the most desirable attribute that their power output can be predicted years in advance. This is not the case with wave energy.

Tidal range generation exploits the potential energy of the difference in the height of the water level at the edge of the oceans that is caused by the tides. A barrage is constructed across the mouth of an estuary to create a difference in water height inside and outside an impounded area as the height of the tide varies. Low head hydro turbines are placed in the barrage. The hydro turbines generate electricity from the potential energy of the difference in the height of the water across the barrage. As the tide rises and falls this difference in height results in flows through the turbines into the impounded area (flood generation) or out of the basin (ebb generation).

Tidal range technology has been studied extensively and is now well understood. Several demonstration schemes up to several hundred MW_e have shown that generating electricity from the tidal range can be technically successful and the future exploitation of this source of energy now depends on its cost-effectiveness and environmental impact.

Tidal stream generation is a quite different technology that uses the kinetic energy of a tidal current to drive a submarine turbine. The operation of a tidal stream turbine has some similarities to that of a wind turbine although a submarine rotor for the same power output is of smaller diameter, as water is 820 times denser than air. The speed of tidal flows varies around the world depending on the behaviour of the tidal wave in the deep oceans. Locally the flows change with the shape of the shoreline and seabed and accelerate around headlands and through narrow channels. As with the wind, a tidal stream is not a simple smooth flow and is likely to contain significant turbulence and eddies as well as variations of flow speed with depth. This creates significant time-varying loads on the tidal stream turbine and its supporting structure. A number of designs of tidal stream turbines have been developed and demonstrated with ratings of each single unit up to ~1 MW$_e$. Some designs use ducts or cowls to accelerate the water flow through the turbine.

There are a number of proposals being developed to install arrays of turbines in fast flowing tidal streams off the west coast of Europe. Tidal stream generation is an emerging area of technology and the most effective turbine architecture has yet to be found. Wind power went through a similar process of refinement of the technology in the 1980s before upwind, three-bladed, variable speed horizontal axis wind turbines became commercially dominant. Tidal stream turbine design has yet to converge on a preferred architecture.

Wave power is a different approach to marine energy conversion in which the power of the oceans' surface waves is captured and converted into electricity. Shoreline, near-shore, and offshore floating wave power devices have all been demonstrated but the majority of the wave energy resource is offshore in deep water. Wave power generators are difficult to build cost-effectively and the extreme forces developed in storms have damaged a number of prototype devices. Personnel access for maintenance, the mooring system as well as the power take-off, and cables all pose particular challenges for floating wave energy converters. The number of wave energy conversion concepts and designs under development is even greater than those for tidal stream and no clear favourite has yet emerged. Although prototype wave power devices are being demonstrated, it is at present the least mature of the marine energy technologies and commercial exploitation of the wave power resource is still some years in the future.

Chapter 6
Future energy systems

Energy for a modern society must be affordable, reliable, and sustainable. These three, often conflicting, objectives can be shown in a triangle known as the energy trilemma (Figure 24). The precise wording of each objective, and the relative importance attached to it, varies over time and between countries with differing political priorities. However, national energy policy makers try to strike a balance and position policy somewhere within this triangle. Energy must be affordable to the poorer members of society in developed countries as well as to those in the developing world. The supply of energy must be reliable in terms of access to energy sources as well as the technical performance of the energy delivery systems. Energy supplies must be sustainable with the environmental impacts minimized, particularly greenhouse gas emissions and impacts on human health.

A continued long-term reliance on carbon (coal) or hydrocarbon (oil and natural gas) based fuels is incompatible with the need to reduce emissions of greenhouse gases. Thus although fossil fuels will continue to play an important role in energy supply for some time, their use will decrease over the next thirty years. Immediate commercial considerations may encourage the development of some fossil fuel energy systems but these assets run the risk of becoming stranded as environmental restrictions increase. Some

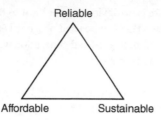

24. Trilemma of energy policy.

large financial institutions have recognized this risk and are already reducing their investment in oil companies and in the generation of electricity from coal.

In many countries only a fraction of the energy used is transported by electricity but its use is likely to increase particularly if cost-effective means of storing it can be found. The only other available energy carriers that do not produce greenhouse gases at their point of use are hydrogen (H_2) or its compounds such as ammonia (NH_3). These can be stored as compressed gases or liquids and when burnt do not produce CO_2 although combusting ammonia will create other potentially harmful gases such as nitrogen oxides (NO_X).

Future electricity generation

Electric power systems can now be operated for hours at a time with 50 per cent of the generation from low carbon energy sources, that is, from variable renewable (wind and solar) and nuclear generation. Until recently, a power system with so much constant output nuclear power and variable renewable generation was considered to be too difficult to control. However, a combination of forecasting the variable renewable resource and provision of reserve generating capacity from controllable renewables (i.e. hydro and biomass), some electrical energy storage, and limited continued use of gas turbine generators has enabled very high levels of low carbon electricity generation to be used. There is no obvious upper limit to the fraction of low carbon generation that can be used but, until energy storage becomes cheaper, some gas-fired generation will be required if costs are not to be excessive.

The drop in the price of PV modules and wind turbines over the last few years means the cost of renewable generation is now at or below 'grid parity', the condition when the cost of generation from an energy source is equal to the retail electricity price. It is questionable whether comparing the cost of renewable generation with the retail price of electricity is appropriate as this ignores all the costs of operating the power system that fall on other forms of generation and the network operators. Typically the cost of generating electricity is only around 30–40 per cent of the retail price with the remainder being made up of reserve capacity, network, and management costs. However, the cost of energy from new utility scale PV and some offshore wind farms (Levelized Cost of Energy) is now reported to be less than the fuel and operating cost of existing fossil generation (Marginal Cost of Generation) in some circumstances (Box 7).

Box 7 Costs of electricity

Electricity systems are expensive and the choice of technologies to be used is heavily influenced by their costs. Unfortunately comparing the costs of electrical energy from different types of generator is not at all straightforward. The elements included in the cost depend on whether the calculation is made by a company considering an investment or to determine national energy policy.

Low carbon generation, i.e. nuclear and renewables, is particularly capital intensive. In recent years the capital costs of wind and solar generation have dropped significantly while in Europe some nuclear plants have experienced cost overruns and delays in commissioning. The time value of money, expressed as the discount rate in a Discounted Cash Flow calculation, influences heavily the apparent cost of electricity from long-lived assets. The costs to the environment and wider society (external costs) are hard to quantify although carbon pricing can go some way to reflect these. The costs of each unit of electrical energy will depend on how much energy is generated (see Box 6, Capacity factor). The capacity factor of a renewable generator depends on the resource, while for all generators it depends on having access to the grid and a market for the electricity.

The Levelized Costs of Energy (LCOE) is often used to make an initial comparison of the costs of alternative electricity generating technologies. In its simplest form, the LCOE is defined as the total costs of the generator over its lifetime divided by the total energy generated.

$$LCOE = \frac{total\ costs\ of\ the\ generator\ over\ its\ life\ (\$)}{total\ electrical\ energy\ generated\ (MWh)}$$

The calculation of the total costs of the generator will be based on a large number of assumptions including capital and operating costs, lifetime, and discount rate. The energy generated assumes a capacity factor. The LCOE is a rather simple measure that neglects the greater value to the power system of generators that have a store of energy and can be controlled to support frequency and voltage.

Future electricity networks and the smart grid

Around fifteen years ago, it was recognized that many of the national electricity networks that had been constructed in the 1950s and 1960s were coming to the end of their lives. At the same time rapid advances were being made in Information and Communications Technology (ICT). Generation powered by renewable energy and low carbon loads such as electric vehicles was also projected to increase rapidly.

The cost of replacing obsolescent networks on a like-for-like basis and continuing with established design and operating practices in the face of these new generators and types of load was predicted to be very high. Concerns were even expressed whether the manufacturing capacity and skilled workforce needed would be available. Thus, there was every incentive to explore innovative alternative approaches to system development both to reduce costs and to increase the functionality of the network. These challenges initiated a period of change in electricity distribution and transmission networks known as the transition to a Smart Grid. Some aspects of this transition are already established while others have only been investigated as demonstration projects.

The fully developed Smart Grid is an electricity supply network that makes much greater use of ICT, including smart meters and real-time control systems. A Smart Grid will use large volumes of data and actively control wide areas of the network, with a greater use of power electronics and the conversion of some circuits to direct current. The Smart Grid will integrate the operation of customers' loads and distributed generation and transform the previously passive distribution network into a continuously monitored and actively controlled system.

Although there will be clear benefits from a smarter grid, significant cyber security risks are anticipated. A key outstanding question is whether the internet can provide suitable security for it to be used for the control of energy systems. As a Smart Grid would increase the power transported through each circuit, there would be an increased risk of instability when faults occur.

Early smart grid initiatives were incremental developments by the traditional electricity utilities, exploiting the advances and cost reductions being made in ICT and addressing the challenges of connecting increasing quantities of renewable distributed generation and low carbon loads. However technology companies and those with a background in ICT (often from Silicon Valley) proposed a number of more radical ideas that challenged the philosophy of the established power system. These innovations were both technical and commercial and continue to be investigated in research and demonstration projects.

Traditional distribution systems were passive networks with equipment once installed left to operate without intervention other than routine inspection or repair after breakdown. There was little monitoring equipment and the network operator had very limited visibility of the state of the network. This is in marked contrast to a modern mobile phone network where the network operator has much greater visibility of the state of the system. In a traditional electricity distribution network, the control systems were simple, being based on technology from an era when communication systems were too expensive to be used widely. Transmission systems, due to their greater importance and fewer elements (i.e. generators, lines, and transformers), have always been more heavily monitored but by expensive and usually bespoke specialist systems. Transmission network control has been local with supervision from manned control rooms and the ICT systems strictly segregated to maintain security.

The emerging Smart Grid has resulted in increased and lower cost monitoring and automatic control of both transmission and distribution systems. However these developments have started from the existing philosophies of operation of each network. There is no fundamental difference between a transmission and distribution system; both transport electricity. In the future there will be convergence both of the operating philosophies and the levels of control and monitoring throughout the networks.

It is traditional practice that once the capacity of a load (or generator) is agreed, a distribution network operator attempts to supply (or accept) the power whenever it is needed. This idea of unconditional access fails to recognize that the capacity of a part of the distribution network to service users is not fixed but depends on the state of the power system and in particular the behaviour of other loads and generators.

As the number of low carbon loads and distributed generators increases and the cost of communication systems reduces, it becomes more cost-effective to integrate the control of loads and generation into the operation of the network. For example a distributed generator could be given permission to connect to the network but only on condition that it would reduce its output when there is insufficient local load. When applied to loads this concept of conditional access is known as Demand Side Integration and for many years large industrial loads have agreed to reduce their demand at times of electricity shortage in return for reduced tariffs. Conditional access to the network can reduce the cost of connecting loads or generators and increases the utilization of network assets. However, it does require monitoring and active control, and increases the risk to the load or generator that they will not be able to receive or generate power for some times in the year.

The increasing integration of small local generation (often in houses) has led to the concept of the Prosumer. This is a customer of the distribution network who both *Pro*duces (generates) and con*sumes* electrical energy. A common example is a domestic dwelling with photovoltaic generation on its roof. During the winter or at other times of low irradiance the Prosumer will draw energy from the distribution network, but around noon on sunny days of low load they will export power. At present it is usually most cost-effective for the Prosumer to use the energy they generate to offset any imports from the network. On sunny days of low load the value of electricity is low and, if the costs are favourable, batteries can be used to store the energy for later use.

The anticipated rapid increase in electrical vehicles poses particular challenges for distribution networks. Although there has not yet been enough experience of electric vehicles in many countries to predict driver behaviour with confidence, one possible scenario causing concern is that many drivers will return home in the evening and plug in their vehicles simultaneously. This would coincide with the existing evening peak of electricity demand. Moreover, it is thought that the owners of electric vehicles may live in clusters and so any simultaneous charging of multiple vehicles would overload the local distribution network.

Smart charging, either through tariffs that change with time or direct control by a network operator, could be used to delay charging until later in the evening when electricity is more abundant and hence cheaper. In some ways an electric vehicle can be considered to be a battery on wheels. Household cars spend more than 90 per cent of the time parked. Thus an obvious approach, which is now being investigated, is to use the batteries of parked electric vehicles as distributed energy storage to support the power system.

Local energy and microgrids

Over the last 100 years, electric power systems have been developed to exploit the technical advantages and economies of scale of large central generators and interconnected high voltage networks. However now there is growing interest in local energy systems, which may be on- or off-grid. In some countries, this interest appears to be stimulated by dissatisfaction with large energy utilities and a desire by some individuals for more control over their energy affairs. In its extreme form this leads to the use of off-grid, dc solar-battery systems that are disconnected from the network. These are becoming popular in some areas with a good solar resource and dwellings with limited electrical loads.

Even when supplied from the main network, local energy organizations are being established to act as energy retailers in an area, buying electricity in bulk and distributing it to customers. Other local energy organizations own and operate their own renewable generators and networks. If more electricity is generated and managed locally, energy use from central generators and network losses are reduced but the need for the wider network to provide back-up generating capacity and control of frequency remains. Traditionally the cost of providing these services was recovered through a charge based on the energy (number of kWh) transmitted through the network. With more energy generated locally these costs will be spread over reduced energy flows, so threatening the business model of the traditional utilities.

Peer-to-peer energy trading is another manifestation of the desire for more local control over electricity supplies. In this model, which is still the subject of research and small demonstration projects, groups of customers that generate as well as consume electricity, and are close geographically, trade energy between themselves and only interact with the wider network as a group. This reduces the electricity supplied by central generators and the participants then hope to reduce their costs. Peer-to-peer energy trading can take place either over the public network or through private wires. One way of organizing peer-to-peer energy trading is to use Distributed Ledger Technology (e.g. Blockchain).

A particularly popular idea is that of a microgrid, although the term is often used loosely to describe a variety of concepts. In its simplest form, it is a collection of distributed generators connected to a section of the distribution network. The voltage and frequency of an on-grid microgrid is taken from the main power system.

Very small microgrids can operate at dc. In this case, batteries provide the stability and voltage reference. These systems are simple and easy to operate but can only supply small amounts of power. Such systems are often used in remote areas of developing countries.

A more radical approach is to operate an ac microgrid that is disconnected, or islanded, from the distribution network. This independent microgrid must provide its own frequency and voltage reference. It also needs some form of energy store to accommodate any mismatch between load and generation. The transition between grid-connected and off-grid operation is technically challenging and if the option of mains electricity is available it is hard to see how off-grid ac microgrids can be cost effective other than in exceptional circumstances.

One undesirable recent development is the increasing use of diesel generators to supply either discrete loads or areas of a city through private wire networks as a microgrid, even though in principle a public electricity network is available. This practice is common in some cities of oil-rich countries (e.g. Baghdad and Lagos) where diesel fuel is cheap but the central electricity generation and transmission system cannot supply the load reliably. Small diesel generators are less efficient and have higher emissions than large central thermal generators and their location in areas of domestic housing creates noise and reduces air quality.

Future gas systems

Pipe systems have been used to distribute gas since the early 19th century, initially carrying manufactured town gas and more recently natural gas. Low-pressure networks serve most of the buildings in many developed towns and cities while high-pressure pipes form transmission systems for bulk transport. Natural gas is mainly methane (CH_4) and so CO_2 is created when it is burnt. Hence natural gas can only continue to be used for some decades or in conjunction with carbon capture if a net zero emission energy system is to be established.

In many countries the gas networks carry more energy than the electricity system. Therefore considerable efforts are being made to identify and trial alternatives to natural gas and so maintain the use of the gas networks. Biogas is made through the biochemical process of anaerobic digestion of biomass. Depending on the biomass, the biogas consists of varying proportions of methane, carbon dioxide, nitrogen, and other trace gases. It can be used to fuel internal combustion engines or burnt directly, but it can also be purified into biomethane by removing the carbon dioxide and nitrogen. It can then be injected into the gas pipeline network. In some countries, financial incentives are offered for biomethane that is injected into the gas grid and small installations are becoming common.

Alternatively, biomass can be gasified through thermochemical processes to form syngas, a mixture of hydrogen and carbon monoxide. The syngas then undergoes a process of catalytic methanation to transform the hydrogen and carbon monoxide into methane and so form bio-synthetic natural gas. The thermochemical transformation of biomass into natural gas is the subject of considerable development activity and could be implemented on a large scale if sufficient biomass feedstock were made available.

Hydrogen is the world's most abundant element and the combustion of hydrogen in oxygen produces only heat and water. The use of hydrogen as an energy carrier has been the subject of research and development for 100 years and indeed coal (town) gas typically contained around 50 per cent hydrogen. Hydrogen can be used to power fuel cell vehicles and to produce electricity and heat through fuel cell CHP plants. There are encouraging demonstration projects using hydrogen to power buses, heavy municipal vehicles, and light trains. In these applications only a limited number of refuelling points are needed.

Pioneering demonstration projects are trialling the conversion of existing gas grids to transport hydrogen but the conversion of the entire gas grid to 100 per cent hydrogen is not imminent. The cost of production of hydrogen from low carbon sources is high and high pressure gas transmission networks contain steel, and hydrogen has an embrittling effect on steel pipes. Current gas appliances are not designed to burn hydrogen and so a reversal of the conversion of consumer gas appliances to natural gas that was undertaken in the 1960s would be required. The public would need to be reassured of the safety of transport and combustion of hydrogen and the current fragmentation of energy utilities may make a programme to convert appliances on the scale required difficult.

A more immediately promising approach is to blend hydrogen generated from renewable energy with natural gas. Countries have different regulations concerning the fraction of hydrogen that can be introduced into the natural gas network. Some of the lower limits, such as 0.1 per cent in the UK compared to up to 12 per cent in continental Europe, appear rather restrictive and trials are under way to investigate the technical limits so that the regulations might be revised. A further alternative is to use captured carbon dioxide in a methanation process to convert the hydrogen generated from renewable energy into a gas that can be injected directly into the gas grid.

Energy Systems Integration

For most of the second half of the 20th century, the supply systems of the different energy carriers were designed and operated largely independently. National electricity or gas utilities commonly concentrated on a single energy carrier in order to build expertise and gain economies of scale. The alternative model of integrating the design and operation of multiple energy vectors in a municipal or regional energy system was adopted in several northern European countries. It is now being recognized that for the development of low carbon energy systems a more integrated and regional approach to energy supply is likely to be preferable.

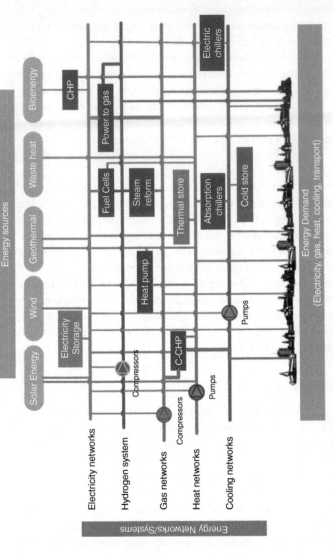

25. **Possible interactions between different energy systems.**

This is known as Energy Systems Integration and has recently emerged as an important topic for research and energy system planners.

The drivers for the integration of various energy systems are to reduce the use of primary energy by making best use of heat rejected from electricity generation, increase the utilization of renewable energy, and provide flexibility for the electrical system while reducing the need for expensive electricity storage.

Examples of the increased coupling between energy carriers include the increased use of natural gas for electricity generation, the increased use of combined heat and power systems, and storage of both heat and cold. Energy storage plays a particularly important role in Energy Systems Integration and while storing electricity remains expensive, it is often much cheaper to store energy as heat or as a gas.

The numerous possible interactions between the various energy systems of a low carbon energy system are shown in Figure 25. The various requirements for energy services can be met from a variety of sources using different energy carriers. The flexibility provided by the energy conversion elements allows the complete energy system to be optimized to minimize capital or operating cost, or emissions.

Concluding remarks

Energy systems are an essential foundation for modern life; they provide great benefit but inflict great damage to the environment. Although encouraging progress is being made, access to clean energy remains out of reach of the poor in some parts of the world. It is widely accepted by scientists and policy makers that energy systems must decarbonize as rapidly as possible in order to limit the emission of CO_2 and other greenhouse gases. The technical and management solutions that are emerging to allow the environmental footprint of energy systems to be reduced over the next 10–15 years are likely to involve: greatly reduced burning of coal and use of oil to power vehicles, increased generation of electricity from renewables, and the replacement of methane with biogas and hydrogen. The route to decarbonize the electricity system is now becoming clear although much work remains to be done. The current pressing challenge is how to reduce emissions from heat supply (particularly for industry) and heavy transport.

Quite naturally, each new energy technology has its enthusiastic supporters but it is naive to expect that a single new technology will meet our energy needs. There are no short-term easy solutions to energy supply and all choices involve compromise. It is hoped that the book has helped readers to understand how energy systems have evolved and has equipped them to take part in the debate about our energy future.

Further reading

Introduction

Maslin M (2014). *Climate change, a very short introduction*, 3rd Edition, Oxford University Press.

IEA World Energy Outlook 2017. <http://www.iea.org/media/weowebsite/2017/Chap1_WEO2017.pdf> Accessed 13 August 2018.

IEA World Energy Outlook Special Report, From poverty to prosperity <https://www.iea.org/publications/freepublications/publication/WEO2017SpecialReport_EnergyAccessOutlook.pdf> Accessed 13 August 2018.

UN Sustainable Development Goals: No 7 Affordable and clean energy <https://www.un.org/sustainabledevelopment/energy/> Accessed 9 September 2010.

IPCC Global warming of 1.5 °C: summary for policymakers <https://www.ipcc.ch/site/assets/uploads/sites/2/2018/07/SR15_SPM_version_stand_alone_LR.pdf> Accessed 18 April 2019.

Lazard (2018). Levelized cost of energy and levelized cost of storage <https://www.lazard.com/perspective/levelized-cost-of-energy-and-levelized-cost-of-storage-2018/> Accessed 18 April 2018.

Chapter 1: Energy systems

BP (2018). Statistical review of world energy <https://www.bp.com/en/global/corporate/energy-economics/statistical-review-of-world-energy.html> Accessed 13 August 2018.

Shephard W and Shephard DW (2013). *Energy studies*, 2nd Edition, Imperial College Press.

Andrews J and Jelley N (2017). *Energy science—principles, technologies and impacts*, 3rd Edition, Oxford University Press.

Fanchi JR and Franchi CJ (2017). *Energy in the 21st century*, 4th Edition, World Scientific Publishing.

Cassady ES and Grossman PZ (1998). *Introduction to energy*, 2nd Edition, Cambridge University Press.

Chapter 2: Fossil fuels

Modern power station practice (1991). 3rd Edition, Pergamon Press.

Miller BG (2011). *Clean coal engineering technology*, Butterworth-Heinemann.

Downey M (2009). *Oil 101*, Wooden Table Press LLC.

American Petroleum Institute. Adventures in energy <http://www.adventuresinenergy.org/index.htmlhttp://www.adventuresinenergy.org/index.html> Accessed 27 September 2018.

Sorrell S. et al. (2009). An assessment of the evidence for a near-term peak in global oil production. UKERC Report ISBN number 1-903144-0-35.

BP (1972). *Gas making and natural gas*, BP Trading Ltd.

Crawley GM (2016). *Fossil fuels, current status and future directions*, World Scientific Publishing.

Helm D (2017). *Burn out, the endgame for fossil fuels*, Yale University Press.

Chapter 3: Electricity systems

Weedy BM et al. (2012). *Electric power systems*, 5th Edition, John Wiley and Sons.

Jenkins N, Ekanayake JB, and Strbac G (2010). *Distributed generation*, IET.

Laithwaite ER and Freris LL (1980). *Electric energy: its generation, transmission and use*, McGraw Hill (UK).

Chapter 4: Nuclear power

Ferguson CD (2011). *Nuclear energy: what everyone needs to know*, Oxford University Press.

Roberts LEJ, Liss PS, and Saunders PAH (1990). *Power generation and the environment*, Oxford University Press.

World Nuclear Association, <http://www.world-nuclear.org/> Accessed 27 September 2018.

Chapter 5: Renewable energy systems

Jenkins N and Ekanayake J (2017). *Renewable energy engineering*, Cambridge University Press.

Peake S (Ed) (2017). *Renewable energy power for a sustainable future*, 4th Edition, Oxford University Press.

IEA (2017). Energy access outlook 2017—from poverty to prosperity, World Energy Outlook Special Report. <https://www.iea.org/publications/freepublications/publication/WEO2017SpecialReport_EnergyAccessOutlook.pdf> Accessed 27 September 2018.

WHO Household air pollution and health, <https://www.who.int/news-room/fact-sheets/detail/household-air-pollution-and-health> Accessed 20 April 2019.

Burton T, Jenkins N, Sharpe D, Bossanyi E (2011). *Wind energy handbook*, John Wiley and Sons.

Kalogirou S (Ed) (2017). *McEvoy's handbook of photovoltaics: fundamentals and applications*, Academic Press.

Greaves D and Iglesias G (Ed) (2018). *Wave and tidal energy*, John Wiley and Sons.

Manwell JF, McGowan JG, and Rogers AL (2009). *Wind energy explained*, 2nd Edition, John Wiley and Sons.

Chapter 6: Future energy systems

Ekanayake J et al. (2012). *Smart Grid: technology and applications*, John Wiley and Sons.

Ofgem, Future Insights Paper 3—Local energy in a transforming energy system. <https://www.ofgem.gov.uk/system/files/docs/2017/01/ofgem_future_insights_series_3_local_energy_final_300117.pdf> Accessed 1 September 2018.

Abeysekera M (2016). Integrated energy systems: An overview of benefits, analysis methods, research gaps and opportunities, Hubnet white paper. <http://www.hubnet.org.uk/filebyid/791/InteEnergySystems.pdf> Accessed 1 September 2019.

Index

A

air quality 13–14
anticline trap 32

B

bioenergy
  anaerobic digestion 109
  biofuel 109
  biogas 109, 126
  biomass 106–10
  sources of bioenergy 107

C

capacity factor 87
carbon capture and storage 30
carbon dioxide 15–16, 30
carbon intensity of electricity
    generation 60–1
climate change 1–2, 14–15,
coal
  boiler 26–7
  composition of 22–3
  electricity generation from 21–30
  mining 19–25
  processing 25
  rank 21
  steel production from 25–6

Combined Cycle Gas Turbine
    (CCGT) 44, 61
Combined Heat and Power
    (CHP) 61
Concentrated Solar Power
    (CSP) 95–6
costs of electricity (LCOE) 118

D

density of energy 17
distributed generation 72
diversity of electricity use 70–1
duck curve 64–6

E

electric vehicles 123
electricity
  alternating current 50
  direct current 49–51, 69
  distribution network 70–3
  frequency, choice of 51
  historical development 49,
    52, 55
  load on GB system 57
  sources of 86
  systems 52–9
  voltage and current analogy 49
  voltage levels 54

energy
    demand reduction 3
    energy and power 6–10
    forms of 7
energy systems integration ESI
        127–9
environmental consequences of
        burning fossil fuels 13–15

F

flow of water in a river 98
flue gas
    desulphurization 29
    from coal fired generation 28–9
    particulates 29
fluidized bed combustor 29–30
fossil fuels 19
fracking 43
future gas systems 125–7

G

gas, future systems 125–7
generations of nuclear reactor
        82–4
generators 52–3
geothermal energy 110–12
GigaWatt-hour 9
glass 93–4
global warming 1–2, 14–15
greenhouse effect 14–15
greenhouse gases 14–15

H

historical use of energy 5–6
Hubbert, M. K. 36–7
HVdc 69
hydrogen 47, 126–7
hydropower
    advantages 96–7
    disadvantages 97
    electricity generation 96–100
    high head 99

I

inertia of power system 67
insolation 89
irradiance 89

K

kilogramme of oil equivalent 9

L

levelized costs of energy
        (LCOE) 118
limiting energy use 6–10
Liquefied Natural Gas (LNG) 45
local energy 123–5
local generation 122

M

marine energy 112–14
MegaWatt-hour 9
merit order of electricity
        generation 63–6
microgrid 123–5

N

natural gas
    composition of 40–1
    formation of 41
    fossil fuel 40–7
    GB National Transmission
        System (NTS) 45–6
    generation of electricity
        from 43–4
    pressure 45–7
    processing 43
    shale gas 41–3
nuclear fuel
    fuel cycle 80–2
    Low Enriched Uranium
        (LEU) 80–1
    mixed oxide fuel (MOX) 81

nuclear power
  Advanced Gas-cooled Reactor
    (AGR) 78, 82–3
  boiling water reactor 78–9
  CANDU reactor 80, 82–3
  electricity generation 3, 74–80
  fast reactor 81–2
  fission 75
  fusion 75
  Magnox 78
  moderator 77, 81–2
  pressurized water reactor 78–9
nuclear reactor
  control rods 77–8
  coolant 78
  generations of 82–4
  heavy water 80
  light water 78
  reactor types 76–80
  small modular 83–4
  thermal reactor 77, 81–2

O

oil
  composition of 31–2, 34–5
  conventional 33
  exploration 33
  formation of 31
  fuel 30–40
  industry 35
  oil field life 12–13, 35–6
  peak oil 13, 36–7
  price of 39
  shale 38
  unconventional 37–8
  well 33

P

parabolic trough solar collector 95
Paris agreement (UNFCCC)
    59–60
power 6–10
Prosumer 122

Q

Quad 9

R

reducing energy use 16
renewable energy
  advantages and disadvantages
    85–6
  electricity generation 85–6
reserves to production ratio 12

S

shale oil 38–9
smart charging of electric
    vehicles 123
Smart Grid 68–9, 119–23
solar constant 88
solar energy
  irradiance 89
  photovoltaics 72–3, 90–3
  renewable energy 88–90
  resource 89
  solar thermal 93–6
solar resource 88–9
storage of energy 17–18
sun 88–9

T

TeraWatt-hour 9
terminology 7
tidal range 113
tidal stream 114
tonne of oil equivalent 9
trilemma of energy policy
    116–17

U

units of energy and
    power 9
uranium, isotopes 77

V

voltage levels 54

W

wave power 114–15
wind energy
    disadvantages of 101

environmental impact 105–6
historical development 100–1
offshore costs 105
power 100–6
wind farms 105
wind turbine 102
world population 1
worldwide use of energy
    10–13

Energy Systems